NOTES on the FLORA

of

NEWFOUNDLAND

M. L. Fernald,

K. M. Wiegand

and others

1925.

Boat = Boat Harbor, Straits of Belle Isle
Bear = Bear Cove (So. of Flower Cove)
Bard = Bard Harbour, St John Bay
Burg = Burgeo & vicinity (Burgeo & La Poile)
Bust = Bustard Cove, St J B —

Eddy = Eddy's Cove. St John Bay

DIRECTORY

Arg=Argentia

Badg Br= Badger Brook
BB= Bonne Bay
B Bulls= Bay Bulls
BC= Birchy Cove (now Curling)
BF= Bishop Falls
BI= Black Island
BJ= Brigus Junct
Ben C= Benoit's Cove and vic.
Bisc B= Biscay Bay, Avalon.
Blom DT= Blomidon, diorite tableland
Blom S=   "   , serpentine tableland
Botw=   Botwood
BP= Birchy Pond (E.Br. of Humber)
BPS= Birchy Pond Stream
Brig= Brig Bay
Big = Big Brook, Straits of Belle Isle
Burn = Burnt Cape, betw Pistolet Bay & Hā-Ha Bay
Brand = Brandy (or Schooner) I., Pistolet Bay

Carb= Carbonear
CH= Cow Head
Chan= Channel
Chim= Chimney Cove
Cl= Clarenville
Cook = Cook's Harbour, Pistolet Bay

DR= Dildo Run (e.end New World I.,etc.)
Don= Donovan's Waterford River
Dead = Deadman Cove (so. of Flower Cove)
Doct = Doctor Hills Bard Harb. Hill.
Ex R= Exploits River
Ed = Eddies Cove, Straits of Belle Isle

FC = Flower Cove
4m = Four-mile Cove, Str. of B. I.
FI = French or Tweed I., Bay of Ids.

Gras = Grassy Island, G. J. Bay

Han . Hannah's Head, Humber R.

Mid = Middle Arm, B. of I.

GeoP= George Pond
GF= Grand Falls
Gl= Glenwood
GP= Goose Ponds
GreenG= Green Gardens, Cape St. Geo.
Gov = Governors I., Bay of I.

Ha = Ha-Ha Mt. & Cape, Straits of B.I.
HB= Hugh's Brook

HH= Hodges Hill
Hol= Holyrood
HR= Harry's River

IB= Ingornachoix Bay
Ice = Ice Point, St. Barbe Bay

KB= Kitty's Brook
Kil= Killigrews

Laz = Lazlade (S.&P.& Miq.)
Lit R= Little River
Look= Lookout Mt., Bonne Bay
Lp= Lewisport
LH = Lark Harbor
L Mt = Lark Mt.

Mauve = Mauve or Noddy Bay, Straits of B.I.
MC= Mistaken Cove, Straits of B.I.
Man= Manuels's River
Marb= Marble Mt., etc. Humber R.
MJ=Millertown Junct.
Musg= Mt. Musgrave
miq = Miquelon

Norris A= Norris Arm
Norm = Cape Norman
No = North Arm, Bay of Islands

OP= Old Perlican
On = Cape Onion & Sacred I.

Sav Id = Savage's Island, Tq Bay

PA= Pikes Arm, New World I., etc.
PaB= Port aux Basques
PaP= Port à Port
Plac= Placentia
PR= Pointe Riche
PS= Port Saunders
PaC = Port au Choix

Q= Quarry
QVL= Quiddy Viddy Lake
Quirp= Quirpon & Quirp. Id.

RenR= Rennie's River
RP= Rushy Pond
Ral= Raleigh & vic., Ha-Ha Bay

Sal=Salmonier
SavC= Savage Cove
SC= Sandy (Poverty) Cove
SEA= Southeast Arm, B.B.
Seem= Mt. Seemore or Steepmore
SL= Sandy Lake
Snook= Snooks Arm
StB= St. Barbe
StBr= Steady Brook Falls
Steph= Stephenville
StG= St. George
StJ= St. Johns
Sum= Summerside, B. of I.
SJI= St John's Island
SI = Seals Nest I., Bay of Ilds.

St P= St Pierre

Tab= Table Mt., P a P.
TC= Tilt Cove
Top= Topsail
Trep= Trepassey
Trin B= Trinity Bay

Virg W= Virginia Water

WB= Western Bay, Conception Bay
WC= Wild Cove, B. of I.
WH= Western Head, New World I.
Whit= Whitbowne
Wo = Woody B., Bay of Isls.
W S = Woods Island

A. M. J. = Anne M. Jeffers

POLYPODIACEAE

Woodsia ilvensis (L.) R. Br.
Carb. PA.TC, Round Harb. GF, StBr.
Flat Bay(Bell) *Doct. On BB*

W. ilv., var. gracilis Lawson (Gaspé)

W. alpina (Bolton) S.F.Gray
TC, Nfd.(Eaton) *Doct. BB*

W. glabella R. Br.
TC, Tab, Nfd (Murray,etc.) GreenG, *Doct*
Han, BB

W. oregana DCEaton (Gaspé)

W. scopulina DCEaton (Gapsé)

Cystopteris bulbifera (L.) Bernh.
Goose Arm (Wagh),Musgr. RiverHead(Wagh),
HR, PaP(JBell), GreenG *Doct Hamb, mi*
BB Romaine Br (RBK).

C. fragilis (L.) Bernh.
Com. Boff, TC BC GF, NA DR Plac,CH Doyles,
Brig. *Save By, Ha Brand, Doct. On Quirp PaC*
BB

C. frag., var. laurentiana Weath. ined.
BC *Doct BB PaP(RBK)*

C. montana (Lam.) Bernh.
  SavC

Pteretix nodulosa (Michx.) Nieuwl.
  Com.W.coast n. to Chimney Cove; Humber
and Exploits Rs. Chimney Cove (Wag), HR,
Romaine Br (Mack & Gr) Doct, BB

Onoclea sensibilis L.
  Com. n. to Notre Dame Bay, Exploits, Humber
and Bay of Ids. HB, Whit, BBulls, Doct Läg, PaC

Thelypteris palustris Schmidel. v. pubescens (Laws.)
  whit. 67 Fern.

T. simulata (Dav.) Nieuwel. (N.S.)

T. noveboracensis (L.) Nieuwl.
  St.J, Don, Top(H&L) By Carb. Cl, White,
  Gr, CairnMt(Bell), Ch H&L), Balena (WP),
PaB, Trep, BBulls. LH Burg S+P. BB

T. Dryopteris (L.) Slosson
Com. Glen, Gr, RomaineBr(H&C), Sqlm(R&S), C.
Cl, Badg, BC, RyansBr, Brig. GreenC. PaB, BB

T. Robertiana (Hoffm.) Slosson
SEA, Tab, Han, MiL, PaC, BB

T. Phegopteris (L.) Slosson
Com Hol, Man, BC, C, Glen, DR Ch, Carb
Salm, Torbay, Balena, GreenG, Burn Burg Laj
PaC, BB

T. fragrans (L.) Nieuwl.(Lab,)

T. frag., var. Hookeriana Fernald
TC, NorrisA, GF, CairnMt (Bell)

T. Filix-mas (L.) Nieuwl.
TC, IB,(LP),Chim(Wag),MiddleArm(Wag),
BC, Benoit Cove,FrenchmCpve(Mack&Grisc)
John's Beach(Wag),RedRocks (C&L),GreenG,
Doet, SJ9, OnDog PaC BB Romanie Br & Black Duck (RB

T. marginalis (L.) Nieuwl. (Gaspé)

T. cristata (L.) Nieuwl.
Whit. Cl. Lpt, NorrisA, BF, GF, HH, GP, IB(LP)
LarkH(Wag),StG(LP) StP, Laj, Eddy

T. X Boottii (Tuckerm.) Nieuwl. (N. S.)

T. spinulosa (O.F.M.) Nieuwl.
Com. Ch (H&L) TC. PaC Eddy
T. n. var x/a

T. spin., var. americana (Fisch.)Weath.
Com,Plac,FrenchmCove(M&G),Torbay,
Balena(WP),Ch (H&L),JunctBr,GF,BC,BJ,
TC,StB, *Oh Doct Mig. Lg. StP BB. Hr (RBK)*

T. spin., var. intermedia (Muhl.) Nieuwl.
Com,PaP,Bof́I(Wag),BJ,Whit. *Doct L Mt BB*

T. spin,, var. fructuosa (Gilb.)
  Com. GF,JunctBr, BofI(H&L),ApseyBeach(Wag)
Doct

Polystichum acrostichoides(Michx.)Schott,
                                    (N.S.)

P. Lonchitis (L.) Roth
  TC,Musg,BC, GreenG, *Burn, Norm. Doct Hh.*
Mid Paule BB

P. mohrioides,var.scopulinum(DCE.)Fern.
                                (Gaspé)

B. Braunii(Spenner) Fee *var. Purshii Fern.*
  Englee(Wag),IB(LP),Chim(Wag),BofI(M&G),
BC,BenoitC, HR,PaP,Tab,StG(LP),RedRocks(C&L)
GreenG *Ha Doct, Oh, BB Romaine Br(RBK)*

Dennstaedria punctilobula (Michx.)Moore
  Balena(WP)

Athyrium acrostichoides (Sw.)Diels
Frenchman's Cove(Mack&Grisc), FlatBay
(Bell).

A. Filix-femina,var.sitchense Rupr.(Caspé)
Doct

A. angustum Willd.
Balena (WP).

A. ang., var. laurentianumButters
Com, BC, StG(H&L), Torbay,Salm(R&S),TC,
GF, VW(R&S),FC,SC,Whit,GreenG Quirp

A. ang., var. rubellum (Gilbert)Butters
Com, n. to NotreDameBay and BonneBay,
SnooksA, BB, BofI(H&L),BlomI,StG(H&L),
GreenG, ~~Doct~~ StG,Big, PaC

A. alpestre(Hoppe)Ryland, ~~var. americanum Bitter~~
Doct

Phyllitis Scolopendrium(L.) Newm. (N.B.)

Asplenium ~~viride Hudson~~ Trichomanes-ramosum L.
GooseArm(Wag),Musg,BC,Tab,GreenG, Big Burn
Doct, Norm, Hau, Mil PaC BB PaP(RM)

A. marinum L.   (NB?)

Woodwardia virginica (L.) Sm. (N.S.)

W. areolata)L.) Moore   (N.S.)

Pellaea densa (Brack.) Hook.   (Gaspé)

Cryptogramma Stelleri (Gmel.) Prantl;
   CH, Musg, GreenG. *Doct. Har, Mid, Pac B/B*

C. acrostichoides R.Br.   (Lab.)

Adiantum pedatum, var, aleuticum Rupr.
   TC(?), Hare Bay(LP), BB, Blom s, *LMr No*

Pteridium aquilinum (L.) Kühn,
   var. latiusculum (Desv.) Underw.
   Com. n, to Notre Dame Bay and Bay of I,
   Sum, Hol (R&S), GF, Ch(H&L),
   Baccalieu I(Sorn,) BC,StG, PaB, *79. B+P*
   B/B

P. aq.,var. pseudo-caudatum Clute
   LP in herb.

Polypodium virginianum L.
   Top (H&L), Hol(R&S), Plac(Wil), *Harbor Main (amce)*
   Baccalieu I (Sornb.) PA, TC, NorrisA,
   GF, CH, BC, Ch(H&L), Balena (WP),
   Gaultois, Ramea, GreenG. *Doct S+P, Lag,*

SCHIZAEACEAE

Schizaea pusilla, Pursh
Trep, BiscB, TC, C, Look, Blom d, Benoit Cove,
StG, PaB, without loc (LP). *Lmt, Burg, Burnt I.*
Indian Bridge (Acad). BB Harrys Br (RBK)

OSMUNDACEAE

Osmunda regalis, var. spectabilis(Willd.)Gray
Com, n. to Notre Dame Bay, Exploits R. and
Humber R.. GF, BF, C, KP, ShoalB, Holy, Tops,
BadgBr, *F9. BC*

O. regalis, var. "pumila" *spect forma nana Fern.*
BB. Blom s.

O. Claytoniana *L.*
Com. n. to Notre Dame Bay, Exploits and
Humber and Bay of I, Frenchm C. Glen, GF,
Blom s, Grand L, Badg Br, Holy. *Doct. F9.*
Burg Lag. BB.

O. cinnamomea L.
Com. n. to Notre Dame Bay, E. Br. of
Humber and Chimney Cove, Whit, Chan, Balena,
GF, BC, PaB, *Doct, Gov, StP. BB.*

OPHIOGLOSSACEAE

Ophioglossum vulgatum L. (N.S.)

Botrychium Lunaria (L.) Sw. ✓
    IB, CH, PaP, FC, StB, GreenG. *SC, Big, Ha, Brad,*
Boat, Com. on Str, Bard, SJd, Quirp, Doct, Han (terminal)
PaC Gra.
    B. Lun v. onondagense (Underw.) Clute

B. simplex E. Hitchc.
    TW, GF, *LH² Curl.*

B. lanceolatum (SGGmel.) Ångstr.
    Nfd. (fide Underw.)
    OK
    B. angustisegmentum
    Brg9

B. matricariaefolium A. Br. (Gaspé)
    OK FC, LH, FJ, SN, Burg Murray's Pnd (And) BB
    Brg9

B. ternatum, var. rutaefolium (A.BR.)
                                    D.C.Eaton
    Barred I(Sornb.), High GregoI(Huntsm.)
    Geo.P, PaP, *LH, FJ, SN, Burg, BB Brg9*

B. virginianum, var. laurentianum Butters
    IB, BR, Marb, StB, Brig. *FC, Mc, Bear, Sdd.*
Brist,

EQUISETACEAE

Equisetum arvense L. *Goose Grass*
    Com. G, Lpt, StG, Rope Cove, *FC, Big, Burn*
Doct. Bard, *StP, Lig. BB Crabbes (RBK)*

E. arv. var. decumbens Meyer
   Man, GF, FC,

E. pratense Ehrh.   (Gaspé)
   Doct, Cull Eddy

E. sylvaticum,var.pauciramosum Milde
   Com. OP, Badg Br, BC, MJ, Fogo, PaB,
   BBulls, Doct, StB, Lag, BB

E. palustre L.
   HB, WC, Marb, GeoP, HR,StG, Doyles(C&L),
   CC

E. litorale Kuhl. (N.S.)
   Lag, Pac

E. limosum, L.
Com. BC, L Red Ind L, PaP, Tops, BadgBr,
   Doyles, RP, FC, Com Lag, BB Crabbes(RBK),
Stephenr (RBK)

E. hyemale, var. affine (Engelm.) A.A.Eaton
   BP,

E. hy. var. pumilum A.A.Eaton   (Gaspé)

E. variegatum Schleich,
   IB, Tao, FC, SC, GreenG, Sav C, Brand Norm.
Boat. Hm. SggPac Crabbes, Barachois Br & Harry Gr (RBK)

E. scirpoides Michx.
   FC, IB,DeGratBay(Huntsm.) SavC, SC,StB,
   Brig, MistC.Big, Com on Str. Sgg,Pac Lag

LYCOPODIACEAE

Lycopodium Selago L.
Com. n. to Notre Dame Bay, Exploits R.
and BB. Snooks A, Holy, Barred I, ChimnC,
TC, Blom s, BB, Carb, HH, Tab, Baccalieu,
Fogo, PaB, Trep, BBulls, *Seem, N a, Bam, Doct
Ho, Quirp, Ort, L Mt.* *StP, PaB, BB Crabbes (RBK)*

L. lucidulum Michx.
Com. n. to Notre Dame Bay, Exploits and
Humber and Bay of I, PaP, Tab, Whit,
Blom, BofI, GeoP, HH, NA, Balena, Lpt,
Snooks *Doct, L Mt. Lag.*

L. luc., var. porophilum (Lloyd&Underw.)
                                    Clute
GP, Benoit C, FG.

L. inundatum L.
StG, Carb, Plac, Glen, CrocH(LP), BF
MaryAnnL, Q, BC, Blom d, St.G, PaB, Trep,
Burg, StP, Crabbes (RBK)

L. inund., var. Bigelovii Tuckerm.
Q, PaB, Trep.

L. annotinum L.
Com. n. to Notre Dame Bay, E. Br. of
Humber, and Bay of I, FranchmC, WB Balena,
Humb, Carb, Chan, ShoalB, Tab, *Quirp, Mig*
BB

L. annot. var. acrifolium Fern.
  GF. BC. *Burg.*

L. annot. var. pungens (LaPyl.) Desv.
  Com. St.G, StJ, Holy, WB, Carb, Q,
  GP, Shoal B, Trep. *Burn.* 4m, Doct. Quirp.
  79, Burg StB, BB

L. clavatum L.   Key to vars.

  a.Most peduncles branched, with
    2 or more spikes b.
    b.Spikes normally 2.
      Spikes sessile or short-stalked,
       short . . . . . . L. clavatum
                    (typical)
      Spikes long-pedicled,
       elongate . . . . .var.laurentianum
    b.Spikes more than 2.
      Spikes 3, approximate,
       short-pedicelled.var.tristachyum
      Spikes 3-5, long-pedicelled,
       the lowest distant
       (2-5 cm.). . . . var.subremotum.
  a.Most peduncles with 1 spike
    Peduncles 0.5-2.5 cm. long:
    spike 1.5-3.5 cm. long
             var. monostachyon.
    Peduncles 4-15 cm. long:
    spike 3.5-11 cm. long
             var. megastachyon.

L. clavatum L. (typical)
  MJ, Q, Glen, Carb, StJ, HR, StG, ShoalB,
  Chah, Trep, BC. *LH PaC BB*

L. clav. var. laurentianum Vict.
  RP, GF,

L. clav., var. tristachyum Hook.

L. clav., var. subremotum Vict.

L. clav., monostachyum Grev. & Hook.
   StJ.

L. clav., var. megastachyon Fern.&Biss.
   St.G,BP,BF,

L. clav. v. brevispicatum Peck.
   mig.

L. obscurum L.
   Com. n. to Notre Dame B., Humb.R.
   and Bay of I, Balena, StG, Humb,
   GF, Holy,Glen, Fogo, Whit,TC, 33 BB

L. obs. var. dendroideum (Michx.) D.C.Eaton
   StJ. StG. StP. Black'l uet (RBK)

L. sabinaefolium Willd.
   HH,

L. sab., var. sitchense(Rupr.)Fern.
   BBulls, Brigus,Carb,Glen, TC, G, BP,
   Seem, GP, Blom s, StG, PaB, BC, BBulls
   & Mt StP. BB

L. alpinum L.   (Gaspé)

L. flabelliforme (Fern.) Blanch.
   StG, StJ, *Laf,*

L. complanatum L.   Key,aftervictorin

   a.Principal stems almost superfifial;
       ascending branches 5-10 cm. long;
       spikes 1.5-3 cm. long b.
   b.Spikes peduncled
           Spikes normally 2-3,with only
               short pedicels..L.complanatum.
           Spikes 1 ....var. canadense.
       b.spike sessile,short,solitary
                   var. pseudo-alpinum.
   a.Principal stems deeply creeping;
       branches 1.5-3.5 dm. long, loose;
       spikes 1-4, on long peduncles
       (5-6 cm.)..........var. elongatum.

L. complanatum (typical)
   TC, Junct Br, StJ, Badger Br, Trep, StG,
Han,

L. compl, var, canadense Vict.
   TC, BC, *79*

L. compl., var. pseudo-alpinum Farwell

L. compl., var. elongatum
   Seem, *Crabbes (ROK)*

L. tristachyum Pursh
TC *near StJohns (Acad), Han,*

SELAGINELLACEAE

Selaginella selaginoides (L.) Link.
Com. FC, Brig. *MJ, Q, QF, E+P, Barred J, Q, SL,
Kil Tc, BC, Musg, Blau's, PS, Tal, Burn, Doct, Quirp,
J9, PaC DB*

ISOETACEAE

Isoetes macrospora Dur.
Lpt. BC, Whit, Torb, PaB.

I. Tuckermani A.Br.
BJ, Whit, *Quirp*

I. hieroglyphica A.A.Eaton
QVL.

I. Braunii Dur.
Kil, StJ, Chan, Blom d, Kitty'sBr, BC,
BF, BB, Q, BP, RP, PaB, Trep. *Burj, 1903*

TAXACEAE
"Palm"
Taxus canadensis Marsh.
Com, BB, Balena, GF, SB NewHarb, StJ,
Badg Br, BoII, Brig. *Doct, Burn, StP, PaC, AB*

PINACEAE

Pinus Strobus L.
   Freq. n. to Notre Dame Bay, Ind.Brook,
   E. Br. of Humber and Bay of I. RP, Blom s,
   Bof I, BadgBr. *No Laŕ. BB*

P. resinosa Ait.
   Terra Nova to Gambo, GP, SL.

Larix laricina (DuRoi) Koch
   Com. BC, GF, FC,Frenchm C, SJ, StG, BBulls,
   FC, *No StP. Laŕ, PaC. BB.*

Picea canadensis (Mill.) BSP.
   Com. WC, BC, Baccalieu I, StJ, GreenG. *Ed
   Laŕ. StP. PaC. BB.*

P. mariana (Mill.) BSP.
   Com. PaP, BC, OP, GF, Shoal B, *Aug.
   Laŕ. Miŕ. StP. PaC. BB.*
   *P. rubra (Du Poi) Koch
      Laŕ.*

Abies balsamea (L.) Mill.
   Com. Chan, Baccalieu I. *OK PaC. BB.*

A. bals., var. phanerolepis Fern,
   Tops, BC, Chan. *Ed, On, StP, Miŕ.*

Juniperus communis, var. montana Ait.
   Com. Fogo, Torbay, PaP, FC, Chan, Sum, CH,
   SL, Seem, DR, GF, FC, CVL, BP, Blom,
   Baccalieu I, FC, SC, BBulls. *Big StP. Mil PaC. BB*

J. com. var. megistocarpa Fern.&St.J.
Blom's, Brig. *Mt* *Murrays Pond (Aud.)*

J. horizontalis Moench *Savin. Face-and-eye Berries*
Com. FC, BB, OP, Glen, DR, GF, Blom,
BF, TC, Marb, Fogo, BarredI, BadgBr,
Torbay, StJ. *Bard, *Mt* S+P, PaC.88,*

TYPHACEAE

Typha latifolia L.
Lpt, Croe Harb(LP), RP. Steph/
BB (abund at Alfalade, harvester for flags in making
barrels, at Baker Brook 12 miles no. of BB (Publ.)

SPARGANIACEAE

Sparganium americanum Nutt.
BJ, StJ, Whit.

S. chlorocarpum Rydb.
Carb, Lpt, BBulls.

S. chl., var. acaule (Beeby) Fernald.
Tops, Kil, Carb, BJ, Whit, Cl, TC, GF,
Chimn Cove, StG.

S. angustifolium Michx.
Com. ChimC, Chan, BluffHd, Kil, QVL,
MiddleCove, Carb, TC, GF C. Look, Blom, BB,
BC, StG, PR, Torbay, Whit, PaB. *On Quirp.*
OtterP, Mt, Brig. *S+P.*

S. multipedunculatum (Morong) Ryob.
B+g. Chimn No

~ S. fluctuans (Morong) Robinson
RP.

S. glomeratum Laestad (Natashy)

~ S. minimum Fries. ✓
TC, GF, RP, BC, GeoP.

~ S. hyperboreum Laestad.
Carb. Whit (Wil), Plac(Wil) Glen ✓
Barred I. Tw. TC, GF, JunctBr, MJ,
StAnth(Huntsm.) FC, BB, Blom d. StG.
Cape Ray. *Norm, SC, Quip, SavC LH, BB*

NAJADACEAE

~ Najas flexilis (Willd.) Rostk. & Schmidt.
BC Buchan(KPJ), Crabbles(RBK)

POTAMOGETONACEAE

~ Potamogeton filiformis Pers.
Plac, TC, IB, HB, WC, Blom s. HR, Geb. ✓
FC, MistC, Brig. *Boat, Bard, Cook, OtterP, S+B*
Pac, BB

P. fil. var. Macounii Morong (Magd.)

~ P. vaginatus Turcz.   (Lab., P.E.I.).
S+B, FC

P. pectinatus L.
   Plac, Arg, NA, PaP, Steph, StG.

P. *zosteriformis Fern,*
~~P. compressus L. (zosterifolius Schum.)~~

P. *Porsildiorum Fern,*
~~P. subsibiricus Hagström~~ (Ungava)

P. confervoides Reichenb.
   StJ, Glen, Junct Br, MJ, Q, PaB. BB

P. ~~Hillii Morong~~ *longiligulatus Fern.*
   FC

P. foliosus Raf.   (Gaspé)

P. Friesii Rupr.
   GeoP. S+B,

P. obtusifolius Mert.&Koch   (Gaspé)

P. *calpophilus Fern*
~~P. rotulosus Hagström~~   (Gaspé)

P. *panormitanus Biv.*
~~P. pusillus L.~~   (Gaspé, Magd.)

P. pus.,var. tenuissimus Mert.&Koch.
   Carb, Whit, IC, BC, Blom d, HR,Steph.
   Ed.

P. *Spirillus Tuck.*
~~P. amorphus Raf.~~
   RP.

P. epihydrus Raf.

Com. n. to Exploits and Humber and Bay
of I, Carb, Kil, GF, QP, BJ, Torb, BadgBr,
BC, FrenchmC, StG, Trep, BBulls. *Burg mij.*

P. epihydr., var. cayugensis (Wieg.)Benn.
(Lab.)

P. alpinus Balb.
Carb, GF, GrandL, BC, GeoP, Steph,
Hermitage(Wag), PaP, FC. *On Quirp BC*

P. pulcher Tuckerm. (N.S.)

P. angustifolius
PaC

P. amplifolius Tuckerm. (Gaspé)
Frenchm Cove B of 9 (m+z.)

P. polygonifolius Pourret
Tops, BJ, PortugalCove, Carb, Whit. *StP,*

P. "natans L."
StJ, Whit, BP, GF, RP, BadgBr, BP, Tab,
Steph, PaP

P. Oakesianus Robbins
Com. n. to Notre Dame Bay, Exploits R.
and Bonne Bay. BB, Blanc, Look, C, MJ,
Gl, HH, TC, GF, PaS. *Quirp On BC*

P. gramineus L.
Com. Bč, PaP, JB, Lp, BJ, RP, Mj, Cl, Stj, Bear
Sav, 7 Cr Pal 8B,

P. gram., var. spathulaeformis Robbins
GP, HS, Brig. oder P. S+B

P. praelongus Wulfen
TC, SP, Whit. Cook Bolo.

P. Richardsonii (Benn.)Rydb. (Gaspé)

P. perfoliatus L. var. graciliones]
Whit. Na

P. bupleuroides Fernald
Tops, Kil, whit, Holy, Na, WC, Stö, PaP.
Mj. BB

P. subnitens Hagström.

Zannichellia palustris L. (Gaspé)

Z. pal., var. major (Boenn.)Koch
Na, Steph, Stö, S+B, BB

Ruppia maritima L., var. obliqua (Schur)
Aschers. & Graebn. (Magd.)
Mj. BB

21

R. marit., var. intermedia (Thedenius)
        Aschers. & Graebn.   (Gaspé)

R. marit., var. brevirostris Agardh (Magd.)

R. marit., var. longipes Hagstrom
    Kil. BB

R. marit., var. rostrata Agardh (Gaspé)
    LH, Burg.

R. marit., var. subcapitata Fern.&Wieg.
                                (Lab.)
    BB

R. marit., var. exigua Fern.&Wieg.
    NA.

Zostera marina L.
    HB, FranchmC. BB

Z. mar., var. angustifolia Hornem.
    DR, Steph, Arg. Ed, Bard, Burg. Lag, BB

JUNCAGINACEAE

Triglochin palustris L.
Com, FC, TW, TC, IB,CH, FrenchmC,RopeC,
StG, GF, BC,Plac. Sgg Burg, PaC BC,

T. maritima L.
Com. on west coast,local eastw. Gl, TW,TC,
BF,GF,Q,FCS,BB,Blom,FrenchmC,IB,Look,Tab,
StG. Burg miq PaC. BB.

Scheuchzeria palustris v. americana Fern.
Gl,DR,GF,HH,MJ,Q,SL,BC,Blom. BB

ALISMACEAE

Sagittaria graminea Michx.
BJ,Whit,GF,RP,HH,BC.

GRAMINEAE

Bromus ciliatus L.
Freq. Tab, PS,GF,GeoP,HR,BB,Brig. On, Burg

B. ?
BB

B. cil. var. denudatus (Wiegand) Fern.
Com. SavC,Brig,TC,GF,Carb,Don,MJ,Musg,BF,
BC,PS,Steph,Tab, StG,Man,Saug,SavC Dot,
miq.

B. inermis Leyss.

RP.

B. *hordeaceus* L.
mig

Bromelica striata (Michx.) Farwell
GI, TC, HH, C, FC, BB, Blom, HR, Tab. *SavC.*
Burn, Doct. & Int. BB.

B. of
pale

Pleuropogon Sabiniei R. Br.   (Arct)

Glyceria melicaria (Michx.) Hubbard

G. canadensis (Michx.) Trin.
Com. n. to Notre Dame, E. Br. of Humber, and
BB, StJ, WB, RP, Garb, Torb, Mary annL, KB,
TC, Tops, Whit, BF, BC, Look, P&P, StG. *& Int.*
StP. Crabbed (RBK)

G. laxa Scribn.   (N.S.)

G. ~~nervata (Willd.) Trin.~~ *striata (Lam.) Hitchc.*
ShoalHarb, GF, OldFerolle, BC, Fr.Cove. *P&C*

G. ~~nerv~~, var. stricta (Scribn) *Drw.*
Com. BB, Whit, GI, GF, RP, MaryAL, BC, LarkH, FC,
StS, SGS *Quirp mig Whit Pal. BE*

G. grandis Watson
BC (E. & G.)

G. Fernaldii (Hitchc.)St.John
QVL,MidCove, GF,BC. *Bard*

G. fluitans (L.) R.Br.
QVL,MidCove,StJ,BBulls, Trep. *BB (Jansson)*

G. borealis(Nash)Batcheléer
Whit,Lpt,JF,GF,KB,CH,StG,Steph,Boyles,
PaB,LitR. *StP, Barachois Br (RBL)*

Puccinellia maritima(Huds.)Parl.   (N.S.)

P. phryganodes (Trin.)Scribn.&Werr. (Lab.)

P. distans (L.)Parl.   (N.S.)

P. coarctata Fern.&Weath. *FC Sg3, Quirp,*
    Torbay,DK,BarredI,FunkI,CH.
On,

P. laurentiana Fern.&Weath. (Gaspé)

P. macra Fern.&Weath.   (Gaspé)

P. paupercula (Homl)Fern.&Weath.
    KiI,NA,FC. *On, Boat, Isth, Bard, Sav, Wo, Bury*
BB

P. paup., var. alaskana (Scribn.&Merr.)
          Fern.&Weath.
OH.StG.FC.LitR. *Brand Pac*

Festuca rubraL.
  Com. Man. QP,Torbay,MJ,Plac,Tw,Fogo,BB
  GF,Gl,GP,Blom,Tab,PA,RP,MistC,Brig,BBulls.
On norm,Doct - StP Pac ,BB

F. rub., var. multiflora (Hoffm.)Aschers.
                    &Graebn.
  StG,SC.*Raleigh, Doct.Cili*

F. rub.,var. arenaria (Osbeck)Fries
    St.J, GoVI. *mid BB*

F.rub.,var, juncea (Hackel)Richter.
  BB,Frenchm Cove. *Earachus(RBK)*

F. rub.,var. prolifera Piper
  Marb. Tab. *Han. mid BB*

F. ovina L.
  GF.

*F. longifolia Thuill.
  Black Duck(RBK)*

F. ov., var. <u>duriuscula</u> (L.) Koch

? F. sciaphila
  Hart. BC

F. supina Schur
  Tab, seem, Burn, Cook, Ha, Ral, Quirp, On, Cook
Mauve

F. vivipara (L.) Sm.
  SC, FG, GoVI, Fogo, Boat, Sgg, Quirp, On, Burnt,
Lmt. FJ. mil Pac

F. capillata Lam.
  GF, TW, Torb, StJ, PaB, HarbBret. Burg.

F. scabrella Torr.
  Blom s.

F. notgar
  BC

F. elatior L.
  Whit. mig

Poa annua L.
  Com. Nil, GF, BC, Franchmo, StJ, FC. Mauve
Burg, StP. Pac

P. compressaL.

  Freq. n. to Exploits R. and Bay of I.
  whit, PlacJ, Gl.

P. —— (caespitose)
  SC, Bg

P. eminens J.S.Presl.
  Quirp.(LP),MistC,IB,CH. *Bič. Bau, On. Ješhm.*
Pač.

P. alpina L.
  NewHarb(W),PA,Croch(Banks),FC,IB,CH.
  Bofi,Tab,FC,GreenG, *com,ou Str.* *Ačt.*
  Haŭ. mid Pač BB.

P. alp., var. Bivonae(Parl.)St.John
  Tab. *Dčt.*

P. *alp.v. frigida (Gaud.) Reichenb.*
  Brig. *Bvát. Biz. Sč. Sgš. Tab. GreenG(Bisc.)*

P. *alp. v. brevifolia Gaud.*
  Bum, Bvát. Savč. Sg9.v Pač

P. laxa Haenke   (Gaspé)
  Dčt.

P. abbreviata R. Br.   (Arct.)

P. Sandbergii Vasey (Gaspé)

P. glauca Vahl.
  Com. Seem, Shoal P. Musg,Chimno,GF,CH, *Dčt,*
  Haŭ. mid 1813v

P. nemoralis L.
  Stš,GF. *Haŭ 1813v*

P. ——
  Bum.

P. ——
  Bum

P. palustris L.
Freq. n. to Notre Dame Bay and Ingorn.
Bay, TC, Snooks, Northern Bight, Tops, Whit,
GL, StJ, GF, PS, IB, StG, FC, StB, BC, *Doct. SN.*
Fac. BB

P. <u>angustifolia</u>

Not worked out

P. pratensis L.
not worked out
Palp. PaB
P. *copisia* ~~il.~~ *rigens Hartm.*   (Gaspé)

P. costata Schum.
Gl, StJ.

P. trivialis L.
Freq. StJ, BC, BB, StB, *Gard, PaB, BB BlackD(RBI)*

P. saltensis Fern.&Wieg. (Gaspé)

P. salt., var. microlepis Fern.&Wieg.
Freq. n. to the Exploits, Humber and
Bay of I. Gl, GF, GP, Musg, StBr, Sun, Tab.
Curl, Humb, BB

Dactylis <u>glomerata</u> L.
Freq. n. to Exploits and BB. BB. Gl, RP,
BC. *StR BB*

Distichlis spicata (L.) Greene   (N.S.).

Colpodium fulvum (Trin.)Griseb. (Arct.)

Dupontia Fischeri R.Br.   (Arct.)

D. micrantha Holm.   (Lab.)

Sieglingia decumbens (L.) Bernh.
    Renк,QV(Will.)StJ, Carb, BBulls.

Catabrosa aquatica (L.) Beauv.
    FC,Brig,TB,Chan,B&B. *Com. ou Str, Bars, Ral, Sav@*
P&C

Cynosurus cristatus L.
    BC. *S+P*

Nardus stricta L.
    Renк.

Lolium perenne L.
    Benton (W). Gr.

L. multiflorum Lam.,var.diminutum Mutel
    Clode Sound

Agropyron repens (L.)
    Com. PS,GI,StJ,FrenchmC. *Miq BE*

A. rep., var. pilosum Scribn.
  Blom.

A. rep., var.   (glaucuas)

A. pungens (Pers.) R&S.   (M.S.)

A. caninum (L.) Beauv.
  NA,BF,GF,Brig,Steph.

A. can., var. tenerum (Vasey) Pease&Moore
  Freq. Kil, Snooks,TC,BF,GF,PS,BB,SEA,BC,
  Tab, LitR. *Isth. Burg. Han*

A. can., var. Hornemanni (Koch) Pease&Moore
  BF,SL,FC, Brig, IB, BB, Blom s, Tab, FC,StB
  Brig,GranG. *599, Burnt, Horn, Burg. FJ, Pic*
  BB

Elymus virginicus L.
  SEA,MidArm. *BB*

E. canadensis L.   (Gaspé)

E. arenarius L., var. villosus E.Meyer
  DR,HareB(LP),SL,Brig,CH,WC,GreenG, *Barb.*
  Ha, Green, FJ, Burg. Pic BB

E. mollis Trin (Lab.)

Hordeum boreale Scribn.&Sm.
  PS *Quirp. Ont. Manuels*

~ N. vulgare L. field, Murray's P (Ama)

H. jubatum L.
   GF(introd.), PS, BayStG, (LP), SandyPt(W). Ho
   BG
Secale cereale L. Murray's P. (Ama)

Trisetum spicatum (L.)Richter,
   var. Maidenii (Gandoger)Fernald
   Tab, FC, GreenG, F9

T. spic., var pilosiglume Fernald
   PA, BLI, SA, TC, Musgr, Tab, Brig, Burn, Boat, Norm,
Doct, SJG, Brand, On & Mt, Lah, PaC BE

T. spic., var. molle (Michx.) Piper
   BF, GF,

T. melicoides (Michx.)Vasey
   BF, GF, FC, BB

T. mel., var. majus (Gray)Hitchc.
   HR, Tab, RomaineBr,

Sphenopholis pallens (Spreng.)Scribn.
   RopeCove(W), PaP, BB Harrys Br (BBx)

Deschampsia flexuosa (L.) Trin.
   Com. n. to Notre Dame Bay, Mt Steepmore
   & Cow Head, PA, Blom, DR, OP, Musgr, Fogo, Bacca-
   lieu I, Tops, StJ, GF, Seem, CH, Chan. Batt, StP.
   BB

Avena sativa L.   A. fatua
   StP. Murray's P (Ama) BB   BB
A. orientalis Murray's P (Ama)

D. flex., var. montana (L.)Ledeb.
ProspectMt CrocHarb (LP),QurponI(LP),
FlC,Brig,PR,PS,Lookout,PaB, *Doct. Quirp.*
Lmt. No.

D. cespitosa (L.) Beauv.
HR,Whit.

D. cesp., var. *littoralis (Reut.) Richter.*
Blom,BBS,Tab,FC, *Norm,*   *ConeR(W), St GeoP.*
GrandL(N). HarrpBr(RBK)

D. cesp, var. *glauca (Hartm.) Lindm.f.*
~~Tab~~,FC, *HarrpBr(RBK)*

D. cesp., var. longiflora
  (D. Bottnica Wahlenb.)

D. alpina   (Lab.)

D. atropurpurea(Wahlenb.)Scheele (Gaspé)
Doct.

Ginannia lanata(L.)Hubbard
Whit,BC(~~Exit~~) *Mig.*

Danthonia spicata (L.) Beauv.
Com, n. to Notre Dame, B.andBB,Holy Look,
Blom,CV,Gl,GF,Maro,Tab,BB,Chan,BadgBr,
StJ,Trep, *SDG'. Burg. No. StP. No.*

D. intermedia Vasey.
~~Blom~~, BBS, *No.*

D. "thermalis Scribn."
   Blom, DG, Mid, BB

stolonifera L.

Agrostis ~~palustris Hudson~~
  Com. n. to Expl. R., and Bay of I. NA,
  GeoP. StP

A. stol. v. major (Gaud.) Fariv.

  stol.   *compacta Hartm.*
A. pal., var. ~~maritima (Lam.)~~
  Com. TC, Fogo, Tops, GF, Sum, CornBr,
  FrenchmC, Chan, Burgeo, Brig. *Bad StP BC*

A. tenuis Sibth.
  Com. n. to N.D.Bay and Ingorn.Bay.
  GF, BB, BC, PaP, Snooks, Carb, Ren, Tops, WB,
  ChimC, Whit, PS, Trep. *Mig BB*

A. hyemalis (Walt.) BSP.
  Com. GF, BadgBr, RP, GF, Carb, PS, Blom, BayBulls,
  WB, GrandL, Tops, BC, StJ. *Bard*

A. hyem., var. elata (Pursh) Fernald
  Com. Tops, BB, BC, BBulls, Trep, Brig. *Mauve,*
Bury,

A. hy., var. geminata (Trin.) Hitchc.
  Com. Sum, StC, Seem, SL, GP, GF, Look, Torb,
  GrandL, Marb, SavC, StB, HarbBrit, Trep. *Ha, Quirp.*
Bard, LH, Bury. BB

A. perennans (Walt.) Tuck.
  LH, LM, Bury,

A. canina L.
  Torbay, StJ, Brigus (R.Bell) Carb, OP, Whit,
  BBulls, PaB.

A. paludosa Scribn.
  SJ, PanC

34

A. can., var. mutica Gaudin.
   Glen.

A. borealis Hartm.
   GrandL(W),FC,PR,BBs,Blom,HR,Tab. Boat, Ha
Quirp. LMt. Doct, BB

A. borealis,v. macrantha Eames
   Blom. LMt.

A. melaleuca Trin.
   FC Doct
A. Rosaae Vasey
   BB (Jamison)
Arctagrostis latifolia(R.Br.)Griseb.   (Lab.

Calamagrostis Pickeringi Gray
   Com. n. to Notre Dame Bay, Exploits,
   Humber and BB, Gl,Fogo,TC,Blom,BP,
   JunctBr,SL,GP,GrandL,Look. Quirp, Bur. LH,
   BB

A. Pick.,var. debilis (Kearney)Fern&Wieg.
   Com. similar range.STJ,WB,Carb,TC,BP,
   GP,RP,BadgBr,SL,BB,BC,GrandL,Blom,
   GovI,PaB,Trep. Or. Bur. Miq

C. canadensis (Michx.)Beauv.
   Freq. n. to ExploitsR. BB,WB,STJ,RP,
   KiI,NA.

C. canad.,var. robusta Vasey
   Com. RP,JunctBr,Whit,BC,HR,Steph,OP,GP,
   PA,Snooks,Blom,SEA,Look,BBulls,STJ,Sum.
Quirp, Doct, Bur,   SPp. Miq

C. can., var. Langsdorfi (Link) Inman
Torb, OP, FC, Brig, SBF, StG. *Sgd.*

C. neglecta (Ehrh.) Gaertn., Meyer & Scherb,
var. borealis (Lange) Kearney
GrandL(W), FC, BofI(W), Tab, LongPt(W) *4m. Burn.*
Quirp, Jth. On ~~*Cas 2)*~~ *BB*

C. scabrous V
BB

C. purpurascens RBR.   (Gaspe')

C. hyperborea Lange
Com. from Exploits W. and N. BF, GF, PS,
SEA, GooseA, GrandL, Musgr, HR, Tab, Brig, FC,
GreenG, *Baid, Doet, OtterP, Gov, LMt, No.*

C. hyp., var. elongata Kearney
Blom s,

?C. Gppon
Pic.

Ammophila breviligulata Fernald
IB, StG, CapeRay, LitR, *Buy, Woody9, Wood9, Gov,
StP, Miq*

Cinna latifolia (Trey.) Griseb.
Com, Snooks, Tops, Plac, PlacJ, GF, BB, BC, PaP,
StG, FC, *Doet, Buy, Lang,*

Muhlenbersia
~~Sporobolus~~ uniflora (Muhl.) *Fern., v. terrae-novae Fern.*
Freq. n. to exploits, Humber and Bay of I.
Caro, StJ, GF, RP, BadgBr, BC, PaB, *LH, Buy,*
BB

Phippsia algida (Soland.) R.Br.   (Lab.)

Alopecurus pratensis L.
   Com. BI,Torb,Sum,SJ,Gl,StB,Trep,*Quirp.*
Cartyville(RBK)

A. geniculatus L.
   Com. ~~a. to~~ Exploits and Bay of I. *Bing.*
   PaB,BC,Trep,
   StP.

A. aequalis Sobol.
   PaP, *Bard, BB*

                        natans (Wahlenb.) Fern.
A. aequalis,var.~~Merrieni (Beal)~~   (Lab.)
   Boat

A. alpinus L.   (Arct.)

Phleum pratense L.
   Com. WB,BC,SJ,Grand L,Sum. *Miq.*

P. alpinum L.
   Quirpon I(LP),FC,StB,Hawkes B.*Quirp com on StG*
Doct, BB

Muhlenbergia racemosa (Michx.) BSP. ✓
   BF,GF,BadgBr,BenoitBr,HR,Steph, *Bard,*
BB

M. Richardsonis (Trin.)Rydb.   (Antic.)

Dilepyrum erectum (Schreb.) Beauv.
Com. Trinity Bay and Notre Dame B. to
BonneB.and B.ofI. RP, BB,BenoitC,
BadgBr. *Indian Bridge (Anna)*

Oryzopsis asperifolia Michx.
Blom,LarkH(W),Tab.*Sgg. FS. Ho. PaC, BB*

O. canadensis(Poir)Torr.
Q,Blom.*L Mt, Burg,*

Milium effusum L.
IB *Doct.*

Cynodon Dactylon (L.) Pers.
mis, Lang,

Spartina Michauxiana Hitchc.
Kil,NA,BF,GoVI(E&G),CapeRay, *Lang, BB*

S. alterniflora Loisel.
StG,Steph'

S. patens (Ait.)Muhl.
StG,

Hierochloa odorata (L.)Wahlenb.
FC,StB,Brig, *Ha, com on Str, Quarp, On, Brig, Lang*
PaC Harrys Br (RBK)

H.od,var. fragrans(Willd.)Richter,
DR, PS, *Burg,*

38

H. alpina (Sw.) R.&S.
  Fogo, Baccalieu1, HH, TC(MissPriest).
  Blom, *Doct, Quirp & Int, F9, StP, BB*

H. pauciflora R.Br.   (Arct.)

Anthoxanthum odoratum L.
  Com. n. to ~~Exploits~~ R. and Bay of I.
  Tops, BB, FranchmC, G1, RP, StJ, BC, *StP*

Phalaris arundinacea L.
  BF, GF, RP, HH,

Echinochloa Crus Galli (L.) Beaw.
  StP.

Panicum boreale Nash.
  Freq.n. to Bay of I, Exploits and Gander,
  RP, GF, BF, BadgBr, G1, BF, GF, Blom.

P. subvillosum Nash (Antic.)

P. Lindheimeri, var. fasciculatum (Torr.) Fern.
  TC.

Setaria lutescens (Weigel) Hubb,
  StP.
  5. viridis. v. Weinmanniana
    murrayi P (Aust.)   CYPERACEAE

Dulichium arundinaceum (L.) Britton
  G1, BF, GF, BadgBr,

Eleocharis ovata (Roth.) R&S.
  RP.

E. palustris (L.)R.& S.
Dact, Norm, StB, PS, Cof B,

E. ~~nigens (Bailey) Fern. ined.~~ *palustris, var. major Sonder*
Whit, Tilt, RP, Look, Curl, Trench C. Bug, Cabbes (RBK)

E. Smallii   (N.S.)

E. *uniglumis*   (~~compressed stem~~) (Gaspé)
Blom

E. ~~halophila Fern. ined.~~ *uniglumis, var. halophila Fern. & Brack.*
Steph, PaP, Bug, mig, StG, BB, Cabbes (RBK)

E. calva
BB

E. acicularis (L.) R.&S.
QVL, MidCove, StJ, Kil, Carb, BJ, Whit, Clar, GF,
RP, KitBr, *Pig, Cabbes (RBK)*

E. acio, var.
BB

E. capitata (L.) R.Br.
Freq. m. to Exploits R and IngornBay. IB.
Gl, BT, Carb, GF, GrandL, StJ, CoalR, ChimnC,
BBulls, *Stg, BB*

E. nigg?
BB

E. nitida Fern.
JunctBr, Look, *BB, millertown (Janson)*

E. intermedia (Muhl.)Schultes   (Gaspé)

Scirpus nanus Spreng.
   NA,StG,Steph. *Quig, BB*

S. pauciflorus Lightf. f.
   NA,BF,GF,FC,Look,BC,BenoitBr, *Es*, *SC.Bard,*
   Pac BB, Bauachois Br, Robinson's Crablees (RBK)
   S. aepinus Schleicher (Ming.)

S. cespitosus L.,v.callosus Bigel.
   Com,StJ,BB,Musg,Balena,MJ,IB,Q,FOGO,
   PaB. *Quirp, StP, Harrys Br (RBK)*

S. cesp., var. delicatulus Fern.
   Gl, *BB*

S. hudsonianus(Michx.)Fern.
   Com. Man,BC,Blom,Carb,BB,WB,FC,Salm,
   GF,RedR,CHay,FrenchmC, *SC, Burn, mig Pac*
   Spruce Br (RBK)

S. rufus (Huds.)Schrad.
   DR,FC,PS,PaP,StG,Steph, *SC. Bard. Ral. SavC.*
   Pac Eddys

S. subterminalis Torr.
   GF,RP,BadgBr,HI,BC,StG,PaB *mig*

S. americanus Pers.
   Kil, NA,HB,StG, *mig BB*

S. validus Vahl.
   HB,

S. acutus Muhl.
   RP,GovI,PaP,Steph, *no*

S. campestris Britton,v.paludosus(A.Nelson)
                        Fernald   (NS)

S. rubrotinctus Fern.
   Com. Gl,GF,Look,PS,BofI,Tops,Chimne,StG,
Or, Doet, Joth. BB, Lang,

S. atrovirens Muhl.,v. georgianus(Harper)
                          Fern.
   TC,GF,WC,

S. cyperinus (L.)Kunth,v. pelius Fernald
   Freq.n. to Exploits R, E. Br.of Humber
   and Bay StGeo. Torb, RP,Q,StG, *Crabbes(RBK)*

S. atrocinctus Fern.
   Com. BB,Q,Don,TC,GF,Musg,Steph,StG,StJ,Whit,
Bard, LH, Crabbes(RBK)

Eriophorum chamissonis C.A.Meyer
   MJ,Q,GaffTops,FC,Blom,BB,FrenchmC, *Brand,* SC,
Rde, Quirp.

A. Scheuchzeri Hoppe
   FC,

E. *Chan, v. aquatile (Norman) Fern.*
   Quirp.

42 *spissum* Fern.

E. ~~callitrix~~ Cham.   *Bog Cotton*
Com. StJ, Bac, Fogo, BaydeNord, MJ, OP,
Musg, Balena, Q,GP, LarkH, Blom, FC, *Quirp*, SC,
    StP. PaC, BB

E. ~~cal.~~ *spis*, var. erubescens Fern.
Com. n. to ND Bay, Exploits and CowHead,
FrenchmC, Q, Blom, FC, MJ, BB, GP, Seem, CH,
Govl, Chan, PaB, *Sav C, Big, Quirp, StP. PaC Doct*

E. callitrix Cham
    Big, Brand, Cook, Boat, SavC
E. opacum (Bjornstr.) Fern.
    FC, SC, SavC, Big, Brand Cook, PaC

E. gracile Koch
    GF, MaryAnnL, BC, Blom, LH

E. tenellum Nutt.
    Com. n. to N.D.Bay, Exploits and BB.
    Q, BC, StJ, Carb, TC, RP, Look, BC, StG, BBulls,
    Bard.

E. angustifolium Roth   *Bog Cotton*
Com. OP, Carb, GF, GrandL, Blom, BB, Chan, DR,
MJ, GP, LarkH, PaB, FC, *Quirp, SC, StP.*

E. viridi-carinatum (Engelm.) Fern.
    Com. Tops, WB, Carb, Gl, MJ, GF, FC, BB, IB, BC,
    FrenchmC, Sum, *SavC, Big, Sa, Sgg Quirp. Plopp*
    Spruce Br (TBX)

E. virginicum L.
  Com. n. to N.D.Bay,Humber and Bayofl.
  OP,Torb,StJ,Carb,Tt,C,RP,Man,Holy,StG,
  BBulls, Look, *Burg St. mig. BC*

Rynchospora fusca (L.) Ait. f.
  Holy,Plac(Wil)Tt,BF,GP,JunctBr,G,Look,BC,
  Burg BB Black Duck (RBK)

R. alba (L.) Vahl
  Com.RP,HB,Look,Ch,Man,Tw,StJ,Carb,GrL,
  Steph,PaB,Trep,*Bard, FI Burg mig BB*

R. capillacea Torr.   (Gaspé)
    var. leviseta EJ.Hél
    BB (Jansen).   *BB (Jansen)*

Mariscus mariscoides (Muhl.)Kuntze   (NS)

Kobresia simpliuscula(Wahlenb.)Mackenz.
    Tab,PS,GreenBard,BC,SC,Brig,MistC,GreenG,
  Big. Burn. Com on Sti. Cook, Boat. 5J9. OR. miĹ. PaC
  BB

K. Bellarsi (All.)Degland

Carex nardina Fries   (Gaspé)
  PaC.

C. pulicaris L. ("Canada".Pursh inGray·
                                    Herb)·

C. capitata L.   (Gaspé)
Big, Brand

C. incurva Lightf.
FC, *Dead, e. to Paynes c*, *Ed. SC, Bear, Green J*
Pac, Grab, Barbi

C. gynocrates Wormsk.
    Freq. s. to NDBay, Exploits R and
    Bay St.Geo. TW, TC, Q, BC, Blom, CStG
    FC, Brig, *Brand, Sgg. Boat, Quirp, LH. Pac*
Pac, Robinsons(RBK)

C. chordorhiza Ehrh.   (Gaspé)

C. Chord. var. sphagnophila Loestad.
    FC, *Ed, SC, Ral.*

C. vulpinoidea Michx.
    TC, GeoP, *Lang.*

C. diandra Schrank
    Freq.on w.coast. FC, PS, PaP, Doyles, GreenG.
Brand, Bard, Pac

C. Paucivaginata (Kutenth.) Mackens.
Salmonier (Ama)

C. stipata Muhl.
Con. n. to NDBay and Ingorn.Bay. Salm.
StJ,Whit,Gl,GF,BC,BB,BC,FrenchmC,PS,StG,
CoalR. Doct, LH SP, BB, Eddy

C. exilis Dewey
Com. n. to NDBay, e.Br.ofHumber and BrigB.
Brig,BC,Doyles,MJ,OP,FrenchmC,GF,Blom,
Holy,Gl,Trep. SavC, SJJ, Quirp SP, mig PaC B8
Bauachms B (RBA), Black 2 (RBA)

C. interior Bailey
Com. s. to GanderR., ExploitsR, and Harry's
R., BB,Gl,TW,GF,FC,PS,CH,BC,StB, Bard, Quirp,
NW

C. Howei Mackenz.   (N.S)

C. sterilis willd.
PS,Brig,Tab,GreenG, 7C, SJJ, PaC

C. Josselynii(Fern.) Mackeny, ined
FrenchmC, BB

C. Wiegandii Mackenz. ined
Freq. n. to NDBay,ExploitsR,and BayofI.
FrenchmC,Q,JunctBr,TC,DR,MaryAL,BC,PaB,
BB

C. incomparta Bickn.   (N.S.)

C. atlantica Bailey   (N.S.)

C. muricata L.
    Blom, *BB*

C. mur.,var. angustata (Carey)
    Com. n. to NDBay, E.Br. of Humb.& Ingorn,
    Bay. BC,Blom,C,GF,Cl,Sum,MJ,CP,GF,
    Sothin, Quirp, On, *BB*

C. mur.,var. cephalantha (Bailey) *Wiegand & Eames*
    Com. n. to NDBay, E.Br. of Humb.&
    Bay,Fl, OP,GF,BJ,Torb,StJ,BF,StG,Blom, *BB*
    Badg,DR,Chan,Fogo,TW. *LH, Mig*

        C."grypo"
        BB

C. Deweyana Schwein.
StG, *Sav, Big. Doct.* *Eddy's* *BB*

C.Dew.,var.collectanea Fern.   (Gaspé)

C. remota L.
(Nfd. acc. to *Gay*)

C. Crawfordii Fernald
    Freq. n. to NDBay, ExploitsR.& IngornBay.
    Gl,MaryAL,PS,TC,BJ,Whit,GF,C,GeoP,BBulls.
    Bard, Doct. *BB*

C. scoparia Schkuhr.
    Freq. n. to NDBay,Expl.R.and Harry'sR.
    GF,NA,TC,Cl,Whit,BJ,ShoalH,GeoP,Cl, *BB*

C. scop., var. subturbinata Fern&Wieg.
BF,GF,RP,

C. projecta Mackenz.
Gl,BF,GF,HH,RP,WC,HR,BB, *Tompkins, Doct,*
Burg. *Eddy*

C. Bebbii Olney
Lpt,GF,BC,

C. tincta Fernald
(acc. to Kukenthal)

C. straminea Willd.   (N.S.)

C. hormathodes Fernald
DH,StG, *Burg,* BP

C. cumulata (Bailey) Mackenz.   (N.S.)

C. silicea Olney
StG,

C. adusta Boott
Near Placent Jct,Gl, BF, GF,RP,MaryAnnL,
SL, *Buchan (KB)*

C. aenea Fernald
Near PlacentJc.Gl,NA, BF, GF,MaryAnnL, BP,
SL,KB,GrandL,

C. leporina L.
  BBulls, QVL, StJ, Topsail, K*l, Carb, Whit, TC,
  BC, BenoitC, StG. *LH* 68.

Var. *argyroglochin* (*Hornem.*) *Koch*
  BB(*Jansson*)[worm]

C. pratensis Drej. (Lab., Gaspé)

C. ~~festiva Dewey~~ *maeloviana* (Gaspé)

C. tenella Schkuhr.
  Com. GF, FrenchmC, MJ, BC, StB, WoodsI, *Ral, mc*
fc. Ed.

C. trisperma Dew
  Com. GF, MJ, DR, Gl, FrenchmC, BC, Sum, StJ,
  Doyles, PaB, Brig, Afg, *Sacc, Ral, Quirp, stp*

C. trisp., var. Billingsii Knight,
  GP, Trep, *LH*.

C. tenuiflora Wahlenb
  Gl, Fogo, BP, GF, MaryAnnBr, MillJ, BC,
  WoodsI, *Quirp, OK,*
C. tn × trisp
  (*Jansson*)

C. ursina Dewey (Arct.)

C. bipartita All. (Gaspé)
Quirp, Doct.

C. glar., var. amphigena Fern.
  FC, StB, Brig, BC, StG, PaB, *MC, SC, Boat, Baed.*
Ral, Burg, fac, Gas.

C. glareosa Wahlenb,
  Quirp.

C. heleonastes Ehrh.   (Ungava)
?Doct.

C. norvegica Willd.
 DR,FC,PS,HB,PaP,Steph,StG, Pal SauC, LH, Burg,
Bnst,

C. canescens L.
 Com. n. to NDBay,EBr,Humber and
BofI.,FrenchmC,MJ,MaryAL,Q,Balena,
Sum,GP,GrL, Outrp, StP,

C. canescens,v.disjunctaFern,
Com. n, to NDBay and CowH. DR,GI,
CH,Chan,Fogo,PaB, GtP, Crabbes(RBK)

C. can., var, subloliacea Loestad.
 GF,MistC,CodRoy,Ch, Doct,

C. brunnescens Poir,
 Com. DR,FC,FrenchmC,CH,BC,MaryAL,GF,
GI,Chan,NHarb,StJ,Baccalieu,Fogo,
DanielC(Trinity),PaB,FC,Trep, Ha, On, Doct

 C. Erun., var. sphaerostachya (Tuckerm) Kükenth.
    Ha, Burg, StP,
C. Backii Boott.   (Gaspé)

C. leptalea Wahlenb.
Com. MJ,Doyles,FrenchmC,BC,LoOk,
StJ,Tops, MC,FC, Brand, OrL, Ed, Doct, StP,

C. obtusata Lilj.
 (Acc. to Kukenthal)

C. supina Wahlenb.   (Arct.)

C. rupestris All.
  SC,FC,Brig, Tab, GreenC, Sauc, Big, Read, Burn.
Com. on Str. SJ9. On, Doct, mid Pac BB

  C. filifolia Nutt.   (Lab.)

C. scirpoidea Michx.
  Freq. or com. BB, Blom, Musgr, TC, FC, CoalR.
  Seem, PS, Tops, Coal R. Brig, Burn. Com. on Str.
  SJ9, Quirp, L Mt. F9. No. mid mid Pac BB

  C. communis Bailey   (Gaspé)

  C. varia Muhl.   (N.S.)

C. novae-angliae Schwein.
  BBulls, Doyles, Balena(WP) LitR. F9, S+P.
Crabbes (R BK)

  C. Peckii Howe   (Gaspé)

C. deflexa Hornem.
  NA,BE,GF,HH,S,Seem,BP,GP,BenoitC,Blom,
  LarkH,StG,PaB,BBulls. Burg, S+P, Doct BB

C. umbellata Schkuhr.
  BBulls, Trep, BF,GF.

  C. tonsa (Fernald)Mackenz.   (N.S.)

C. pedunculata Muhl.
 Blom, Tab, Doct, Sgg, FJ, Lang, Pac, Bl.

C. concinna R.Br.
 GGard, Burn, Pac.

C. glacialis Mackenz. Sauc, Ice, Dead, Bg, Burn, Com. on Sk,
 FC, Brig, Tab, GGard
Boat, Sgs, Quirp, Pac

C. eburnea Boott
 TC, Marb, PaP, Tab, GGard, Burn, Doct, Holn, Mid
 Pac Bl

C. bicolor All.
 IB, SC, FC, Mc, Brand, Cook, Brat, 4M, Bear, Sgg.
 Pac

C. Hassei Bailey   (Gaspé)
 Pac

C. aurea Nutt.
 Com. on w. coast. CH, PS, Marb, Tab, PaP, Blom,
 ShoalP, FC, Baird, Sgg, LH, Hal BC, Harup Bi (RBH)
 C. livida (Wahlenb.) Willd. (var. typica Dn.)
  Burn.
 C. livida (Wahlenb.) Willd. var. Grayana (Dew) Dn.
  Com. FC, Twil, MilJ, BC, Carb, Glen, FrenchmC, GF,
  GP, Holy, Sauc, &b, Pac

 C. liv. var. rufinaeformis Fern.
  4-Mi

C. vaginata Tausch.
 Freq. s. to B St G. Exploits R and Gander R.
 FC, GF, Gl, Tab, BC, Burn, Bear, LMt, Pac BC

C. leptonervia Fern.
 Freq. n. to ND Bay, E. Br, Humb, and Bay of I. FrenchmC,
 MJ, BC, BadgB, GF, Tab, Gl, Ha, Bear, Quirp, Curl

C. katahdinensis Fernald
  GF,RP,

C. conoidea Schkuhr,
  Plac(Wil), BF, StP,

C. gracillima Schwein.
  Freq. n. to NDBay,ExploitsR and BayofL.
GF,Blom,CoalR,Steph,Gl, Lang, BB

C. grac.,var. humilis Bailey
  NA,

C. grac.,var. macerrima Fern.&Wieg.
  YorkH,GovI. BB

C. debilis Michx.,var.Rudgei Bailey
  Com. n. to Gander and Expl.Riversand.BB.
FrenchmC, OP,Gl,RedR,Q,BF,MaryAnnL,Musgr.
Look,SL,StJ,DeerL,Blom,StG, Burg Lang

C. arctata Boott
  NA,TC,BB,Blom,Tab,GreenG, BB

C. castanea Wahlenb.
  Gl,TC,BF,GF,RP,MaryAnnL,BP,FC,BC,Blom,
HR,Tab.SavC,Bear, BB

C. capillaris L.
Com. on w. coast. freq. e.to ND Bay and
Gander R. Gl, CoalR, TW, TC, FrenchmC, CH, Musgr,
BC, GooseA, PS, FC, Brig, CRBy. *Mauve, Humb*
Pac BB, Crabbes & Robinsons (RBK)

C. atrofusca Schkuhr. (Lab.)

C. misandra R.Br. (Arct.)

C. misandroides Fern.
Tab, GGard,

C. pallescens L.
3Bulls, Gl, GF, MaryAnnL, Sum, BC, HR, Steph, StG,
LH, Loring, BB

C. scabrata Schwein. (Gaspé)

C. rariflora (Wahlenb.) Sm
Freq. near w. coast. Litk, IB, GovI, CH, FC,
Blom, CStG, PaB, FC, *Brand, Quirp, LH, Burg,*
Miq, StP, Pac BB

C. limosa L.
Com. Q, TC, MJ, Blom, Gl, GP, Look, GF, PaB, FC, *Sgg,*
Lang, Pac BB

C. paupercula Michx.
Com. Fogo, WB, FrenchmC, BB, DR, Gl, MJ, GF, Sum,
StJ, *Quirp, On, Burg, Miq, Look, Pac Hamps (RBK)*

C. lim x var,
Pac BB

C. paup., var. brevisquama Fern.(Lab)

C. Buxbaumii Wahlenb.
Com.FC,GF,MJ,Q,GI,Whit,Carb,BC,Look,IB,SC.
Brand. Burin, Ed. Norm. Lary, PaB, BB, Harmbr(RBK)

C. Bux var (white)
BB

C. **atratiformis** Britton
Freq.on w.coast and N.D.Bay.GreenG,NA,Tab,
WC,FC,Doyles,TC,HR,Sampson'sI,(Exploits)Ha,
Quirp, Doct, Butn. BB

× C. virigonensis Fern.
Quirp
C. Halleri Gunn.   (Lab.,Gaspé)
Quirp, On, Mauric.

C. stylosa C.A.Meyer
PaB, Doct, Quirp, On, BB

C. ~~scopor N.H.~~ *rigida Gooden.*
Green'sHarb.(Wag.),Blom,PaB,BB,Doct, Green D,

C. scopulorum Holm
C, Blom

C. acuta L.   (C.Goodenowii)
Com. Tops,StJ,BC,Carb,Q,GI,DR,TW,Kil,Lp,
BJ,Whit,WB,Baccalieu,GreenHarb,PS,Torb,Quirp,
Quirp. Migt. Bac.BB
C. acuta v. strictiformis (Bailey)
Migt BB
C. lenticularis Michx.
Com, n.toNDBay,ExploitsR.and BayofI.
FrenchmC,GI, Doct, BB

C. Haydeni Dew.
StJ.   × C. aabbit
Doct

C. lentic.,var. Blakei Dewey
'Little Red IndL,

C. lentic.,var.albi-montana Dewey
Doct,

C. lentic.,var.eucycla Fern.&Wieg.
BPS,

C. aquatilis Wahlenb.
Com. on w. coast, e. to Exploits and Gander.
Gl,Steph,PS,BenoitC,HR,GF,BB,FC Quirp,SJS,Bea,
mig, Pac BB

C. subspathacea Wormsk. (Lab.Gaspé)

C. salina Wahlenb.
MistC, Boat, SandyC.

C. sal.,var.kattegatensis(Fries)Almq.
NewHarb(Wag),FC,SavC,PS,StG,CRay,LitR,
Steph,CapeRay,CH, Quirp, On, Ioth, Gov, mig
Pac

C. sal., var. pseudofilipendula Kükenth.
On, Pac BB

C. maritima O.F.Muell.
Freq.DR,PS,Chan,FC, Ioth, Quirp, Gov, Brig, SN,
BB

C. crinita Lam.
LH mig,

C. crinita Lam.,var.gynandra(Schwein.)
Schwein. & Torr.
Com. n.toExploits and Humber.FrenchmC,Gl,
BF,DoR,GF,BPS,MarB,Tops,Salm,StJ,Torb,Man,
mig, Crabbes(RBK)

C. Lyngbyei (Lab.)

C. lasiocarpa Ehrh.
Com. n. to NDBay, E.Br.of Humber and
Bay of I. BB,FrenchmC,Doyles,Gl,RP,GP,
BC,Crabb's. LH Eddy

C. Houghtonii Torr.
BF,GF,MaryAnnBr.

C. Hostiana DC., var.laurentianaFern&Wieg.
Brig,HB,Tab. Lang, Pac BB

C.Oederi Retz.
StJ,Tops,Carb,BC,FC,BBulls, StP, Lang

C. Oed., var.subglobosa(Mielich.)Richter
PH, S99 Pac

C. Oed., v. pumila(Coss.&Germ.)Fernald
Com. StJ, Torb, BarredI,RP,Tw,Fogo,IB,
GF,MaryAL,Holy,Carb,FrenchmC,Lot,Gl,
TC,Chan,CH,BC,StG,BadgrBr,SC,BBulls, OH
Ed, Norm, LH, Burg, Pac BW Crabbes(RBK)

C. lepidocarpa Tausch
FC,SavC,Brig,Tab,HR,LitR,StB, Ed. S99
Lang, Humb, Pac, Eddy

C. flava L.
Com.BC,IB,BF,GF,Tops,FrenchmC,Carb,Man,
GF,Snooks,BB,WB,ShoalP,StG,FC. S99, OH, Quirp
Mauve, LH BB

C. fl..var. gaspensis Fern.
Tab,HR,GeoP,GreenC

C. cryptolepis Mackenz.
GF,BP,Kil,WB,Lpt,Carb,BadgBr, FJ,BB

C.
BB

C. microglochin Wahlenb
FC,SavC,StB.Brig,Big,Brand Com. on StG, Ed.
Baid,SJJ,Ral,Boat,Cook, Pac,

C. pauciflora Lightf.
Com.n, to NDB, Exploits,Humb.and Bayof I.
Sun,Whit,GF,Q,MJ,DR,PaB,FrenchmC,Balena
Quirp, BB

C. Michauxiana Boeckl.
Com.n.to N.D.Bay Exploits, Humber and
BonneB.BB,Carb,MaryvAL,Q,TW,MJ,RP,BC,WB,French
mC,GooseP,Badg,StJ,BBulls, mig BB Harry's Br (RBK)

C. folliculata L.
Com.n. to GamboR.,Upper Exploits,E.Br.
Humber and BayStGeo. OP,StJ,Q,GP,Whit,
Steph,PaB, Burg, mig,

C. oligosperma Michx.
Com. n. to NDBay,Exploits and BB. OP,GF
Carb,Gl,BB,RP,BC,GF,Doyle's,Sun,Blom,StG,
Look,GrandL,Badg,Holy,PaB,Trep, Burg mig BB
Harry's Br (RBK)

C. saxatilis L.   (Lab.)

C. miliaris Michx.
  Com. n.to NDBay,Exploits and IngornB,
GP,WB,FrenchmC,BC,PR,Blom,CoalR,Carb,Tw,
GP,Doyle's,BP,GeoP,Big,On,No,Pas BB

C. mil., var. major Bailey (C.rhomalea)
  Q,Whit,GP,BB,Blom,KppeC,SC,R&C

C.Xmainensis Porter
  Com.with C. mil. Badg,Gl,BF,MaryABr,
GP,BC,RP,GreenG

C. vesicaria L.
  Gl,MaryABr,Cl,BF.

C. ves.,var.monile(Tuckerm.)Fern.
  GF,

C. ves.,var.jejuna Fern.
  Gl,RP,

C.ves.,var.distenta Fries.
  SL,GrandL,Badg,

C. ves. v. dichroa Andersss. (Antic.)

C. Graßana?
  Doct.
     C. Colgrata?

C. ves.,var.Raeana (Boott)Fern.
Musgr. *Ok,*

C. rotundata Wahlenb. (Greenl.)

C. membranopacta Bailey (Lab.)

C. rostrata Stokes.
Com. n. to MtBay & Ingorn.B. StG,BF.
Steph,GF,PS,BC,C,JunctBr,Gl,RP,TC,GF,
FrenchmC,Badg,StJ,BB,Trep *Quip, Lff Lang,*

C. rostr.,var.utriculata (Boott) *Bailey*
LitRedIndL,GF,RP,Whit,Badg. *mig,*

C. rostr.var ambigans Fern.

C. Pseudo-Cyperus L.
BF, *Crabbes (RBK)*

C. intumescens Rudge
Cl,Gl,BF,GF,MaryAnnBr,BadgBr,SL, *F3, Burg*
mig Harris Br (RBK)

C. int.,var. Fernaldii Bailey
Plac,Gl.

ARACEAE

Calla palustris L.
  GF, BadgerBr, StG, Steph.

Lemnaceae

Lemna minor L.
    B&B.

ERIOCAULACEAE

Eriocaulon septangulare With.
  Com. n. to NDBay, E. Br. of Humber & Bonne Bay.
  Badg, Salm, Torb, QVL, BarredI, Fogo, TC, Carb, BC,
  Look, Chan, PaB, Trep. *&Mt, Burg, S&P.*

XYRIDACEAE

Xyris montana Ries.
  Holy, Plac(Wil), LittleHarb, BofI, StG, PaB,
&H, Gov, Burg

JUNCACEAE

Juncus bufonius L.
  Com. TC, GF, StJ, QVL, Chan, IB, BB, BC, HB,
  FunkI, Brig, FrenchmC, *Burg S&P. mig.*

J. buf., v. halophilus, Buchen.& Fernald
  KiI, NA, Conche(LB), PS, StG, Steph, Brig, MistC,
Burg.

J. trifidus L.
  Holyr, QuirponI(LP), HH, Seem, Look, Blom, BB,
Doct, Quirp. On, &Mt, F9, Nb.

J. Gerardi Loisel.
  Kil, BC, BB

J. tenuis Willd.
  Com.n.toNDBay, EBrHumber.and BB, BB,
  StJ, GF, Cl, BJ, BF, MJ, Trep, Carb, Whit,

J. Dudleyi Wieg.
  HB, *Gov, BB*

J. Vaseyi Engelm.   (Lah.)

J. setaceous Rostk.
  "Terre-Neuve" acc.to Laharpe.

J. effusus L.
  Kil, Cl, Whit.

J. eff., var. compactus Lej.&Court.
  Com.n.toNDBay, EBrHumb and BofI. Steph,
  C, GF, NA, Wat, Kil, Carb, Chan, Cl, BC, TC, StJ,
  Pops, StG, Trep, BBulls. *LH, Burg, CB*

J. eff., v.conglomeratus (L.)Engelm.
  StJ, WatBr, Kil, BJ, Carb, WB, Whit, PlacJ,
  Plac(Will), NA, *LH, Miq.*

J. eff., var. decipiens Buch.
  BJ,Plac.

J. eff., var. solutus Fern.&Wieg.
  NA,RP,GeoP. *Lay,*

J. eff., var. Pylaei(Laharpe)Fern.&Wieg. ✓
  BrigJ,Whit,Salm,Cl, BF, GF,RP,WC,GeoP, PaP,
  Döet GB,

J. filiformis L.
  Com. n.toN.D.Bay and Bofl. FrenchmC,RP,
  Doyle's,Don,TW,MaryAL,RP,GF,Cl,GeoP,Fogo,
  Badg,PaB,BBulls. *Quirp, Madoc, Buy GB*

J. balticus Willd., var.Littoralis Engelm.
  Com.near coast and inland as an adventive.
  BB,BC,Chan,Blom,RP,FC,StB, *SavC, Gov,Buy, Mig,*

J. balt.,var.stenocarpus Buch.& Fern.(Gaspe)

J. balt.,var.melanogenus Fern.&Wieg.(Lab.)
  J. arcticus. (Lab.)

J. brevicaudatus (Engelm.)Fern.
  Com.StJ,TC,Q,WB,RP,BJ, BC,Blom,PaP,Look,HB, ✓
  BB,Chan,Trep,Brig,StJ. *Manne, Doct, StB, Buy, RH*
  GB

J. canadensis J.Gay.
   Com. n. to MDBay and Bonne Bay. GF,OP
C,BC,Holy,Carb,Blom,BB,PaB,BBulls, LH
BB

J. pelocarpus E. Meyer
   Com. n. to MDBay and BonneBay. StJ, Carb
RP,Look,HB,StG,Steph,Trep,BBulls,PaB, LH'burg
Crables(RBX)

J. subtilis E.Meyer
   GF,RP,HB,GeoP,StG,Steph,

J. militaris Bigel.
   RP,C,s.of Tor'sCove,Avalon and s.of 34 mi.
post--Caplin Bay(Harperized).

J. bulbosus L.
   StJ, Kil, BrigJ,HarbGrace(Wil),Carb,Whit,WB,
Plac(Wil),Trep,BBulls, StP

J. nodosus L.
   PaP,HR,GeoP,Tab, BB

J. acuminatus Michx.   (N.S.)

J. scirpoides Lam.
   Nfd(LaPyl.acc. to Laharpe)

x notans
BB

J. alpinus Vill.
Bf, GF, FC, IB, BB, Blom, Frenchmc, Quirp.
Brist. Ccabbus(RBk)

J. alp., var. insignis Fries.
MistC, Brig, IB, HB, WC, BC, GeoP, CoalR,
Tab, Steph, FC, CoalR, Gov, BB, *Ccabbus(RBk)*

J. alp., var. uni, biceps Laestad.
SJJ, Burn, Cook
J. articulatus L.
Com.n. to N.DBay and BgfI, Steph, StJ
Carb, BB, Don, TW, BI, TC, BC, Sum, Trep. *LH*
BB

J. art., v. nigritellus (Don) Druce
49, BB

J. art., var. obtusatus Engelm. (N.S)
miq. BB

J. biglumis L.
Nfd. acc. to Laharpe

J. albescens (Lange) Fern.
SC, MistC, FC, IB, BB, Tab, GreenC, Brig. *Eto Bay*
Burn, Brand, Cook. SJJ. Quirp. MC, norm, SavC On
No, P.C., Chrys.

J. stygius L., v. americanus Buchenau
Holf, GF, RP, MaryAnnL, C, FC, LoOk, BC, Blom,
StC, Steph. *Quirp EtP, RBk BB*

J. castaneus Sm.
Nfd. acc. to Buchenau *& Engelm.*

J.longistylis Torr.
　BayofI,Overfalls(Codroy) *Buadoisk(RBK)*

J.

Luzula saltensis Fern.　(Miquelon)
　Lang.

L. parviflora(Ehrh.)Desv,
　Com.near W. coast.Balena,Doyles,FC,HawkesB,
PS,StB,*Quirp.Mauve PaB, BB*

L. spicata (L.) DC.
　Tops,Stanthony,*Burn,Ha,Cook,Doet,Quirp,Bn,*
Mauve 79 BB

L. confusa Lindeb.　　(Gaspé)

L. nivalis(Loestad.)Beurl.　(Arct.)

L. campestris(L.)DC,v.comosa(Meyer)Fern&Wieg.
　Whit,Glen,Baccalieu,BC,

L. campestris,v.congesta(Thuill.)Meyer
　PaB, *StP,*

L. camp,,v.acadiensis Fern.
　GF, *StP. Cuhl BlackDuck(RBK)*

L.camp.,v.multiflora(Ehrh.)Celak.
Com.n,toExploitsandBayofI, StJ,Whit,Plac,
BrigJ,Carb,GF,MaryAL,MJ,BC,Chan,BBulls *StP, BB*
Black Duck(RBK)

L. camp., var. frigida Buchenau.
  Com. Qf,StJ,NewHarb,FC,CarP,TopS,FrenchmC,
Glen,Kil,Tw,Quirpon,PS,BC,Trep *Quirp,On,Cook,*
Ha, SN, Pac

L. camp.,var.pallescens Wahlenb.
  Glen.

L. camp.,var. alpina Gawd.   (Gaspé)
  BB

LILIACEAE

Tofieldia minima(Hill)Druce
  Com.near w.coast,s to B.StG. Tab,GGard,
Blom,IB,FC,Brig. *Boat. Sgg, Quirp, On, Cook*
mid Pac BB

T. coccinea Richards.   (Arct.)

T. glutinosa (Michx.) Pers.
  Com. near w. coast and e. to Gander and
NDBay, Holy,IB,BC,GooseA,Blom,Look,Tab,
BP,BB,FC,BF,Gl,PC,GF,*S. Sab, Sgg, Quirp.Cook,*
L Mt mid Pac, Ku (aug)

Zigadenus elegans Pursh   (Gaspé)

Ornithogalum umbellatum.
  Springsdale (aug)

Allium Schoenoprasum L. v. ~~sibiricum (L.)~~ *laurentianum* Fern.
Hartm.
FC, PR, Sav, Sgg, Doct, Ed, PaC, BB

Clintonia borealis (Ait.) Raf. *Poison Berries*
Com. OP, Balena, FrenchmC, Croque, LewisH, BB, C,
GF, MJ, Badg, StJ, StG, StB, Trep, Doct, Bay, StP, PaC

Smilacina stellata (L.) Desf.
Com. on the coast. GGard, IB, FC, CH, Torb, Tab,
Baccalieu, BarredI, ShoalP, StB. Big, Doct, On, FS, Burg
176   StP.

S. trifolia (L.) Desf. ~~Hyacinth~~ *Scurvy Berries*
Com. GI, MJ, GF, Balena, FC, RP, GrandL, Barred I,
GI, GF, Whit, Croque, PaB, Brig, FrenchmC, Burg, StP.

Maianthemum canadense Desf. ~~Malane Berry~~ *Scurvy Berries*
Com. BofI, Blom, Tab, StJ, Baccalieu, MJ, Brig,
StP, BB

Streptopus amplexifolius (L.)DC.
Freq. S, GF, Torb, Baccalieu, GF, BB, Croque, Man,
BofI, StB. StP.

S. roseus Michx.
Freq. BB, Blom, BofI, Tab, PaB, BC, FC, StB. MC, Lug
Mig, PaC
S. oreopolus Fern.
Ha, Brand, Doct.
Trillium cernuum L.
BF, GF, BPS, Humb, BC, GeoP, FlatB(Bell) BB

IRIDACEAE

Iris versicolor L.
  Com. Chan,Sum,GF,FrenchmC,StJ,BiscB,Bear
On,SavC,FC,Baie,bid,BB,PaC

I.setosa,var.canadensis Foster.
  Com.oncoast.FC,CH,Cl,Baccalieu,Barred I,
Steph,DK,Arg,Burgeo. LH SFP. PaC BB,

I. pseudocorus L.
  QVL,

I.tenax Dougl.
  Nfd(Dr. Morrison acc. to Hook.)

Sisyrinchium angustifolium Mill.
  Com. n. to NDBay and IngornB. FrenchmC,Fogo
BarredI,PS,BC,CH,GF,Cl,TC,Tops,PaP,PR,OP,
BBulls,Sum. SJJ,LH,SFP. PaC BBU

S. montanum Greene   (Gaspé)

S. septentrionale Bickn.
  Tab,GreenG,

S. gramineum Curtis
  Salm,Gl,

C. parviflo... ORCHIDACEAE
Bof3 (nap) HBR K&L Salisb

Cypripedium parviflorum Salisb var. planipetalum Fern.
  Com. near w.coast. Blom, Tab, Marb, CoalR,
FC, Brig, LewisH, GGard. Sav, Ice, Blf, Burn, Boat, SJJ,
LMt, No, mia PåC BB

C. reginae Walt.
  Marb, Humb, BenoitC, GeoP, PaB, (C.&L.)GGard,
79. North Harbour, Bay 4 Expl (Blanche Turner)
  printed by Rouleau
  C. passerinum Richardson (Wing.)

C. acaule Ait.
  StJ, NewHarb(Wag), SmithSound(Cormack), Tab,
StG, Balena PaB, StP, BB

  Orchis rotundifolia Banks
    SC, Blf, Burn, Cook, Brand, Norm, Boat.

Habenaria viridis (L.) R.Br var. interjecta Fern.
  FC, SC, MC, SavC, Ice, PaynesC, Blf, Burn, Cook, Norm,
Brand, Boat, 4Mi, Nauk.

                viridis, var
H. bracteata (Muhl.) Gray
  Mainland opp. High GregoI, NDB(Huntsman).
DoCt. BB

H. hyperborea (L.) R.Br.
  Freq. TC, GF, CH, Marb, BC, FC, Sav Burn, Ice, SJJ,
PåC, BB

                              Water Lily
H. dilatata (Pursh) Gray Scout Bottle Smell Bottle.
  Com. BofI, FC, BB, JunetBr, Gl, Don, Carb, Blom,
GF, OP, CH, BC, Tops, Doyle's, LarkH, Salm, Badg,
ShoalB, Fogo, FC. DoCt, FJ mig.

H. clavellata (Michx.) Sprong.
  Com.n. to NDBay and BayofI, Trep, WB, PåP,
Doyle's, Tops, GF, BC, Man, GF, StJ, BJ, Kil, GL, Ci,
ShoalB, Barred I. LH, Burp, mig BB

H. straminea Fern.
Burn, Norm, Brand, 4MiC, SavC, Cook, PåC Burn (Aug)

H. unalascensis (Spreng.)Wats.   (Antic.)

H. obtusata (Pursh) Richardson
  Com. Doyle's,FC,PS,DanielsCove, Trinity)DR,
Blom,CoalR,Sum,PR,CH,SC,FC. *Lang. Pac*

H. obtus. v. collectanea Fern.
  FC,S99. Big. On Pac
H. orbiculata (Pursh)Torr.
NewHarb(Wag.),BofI(E.&G.),BenoitC,Tab,StG,
Steph,(LaPyl.)*L.M. F9 mig HB(RBK)*

H. macrophylla Goldie.
  Whit.

H. n.sp. "bifera"   *H. diphylla*
  Pac.   *B.O.*

H. blephariglottis (Willd.) Hook.
Freq.n.toTrinity.B.and BofI. Holy,BO,PaP,
Chan,PaB,Salm,BiscB.*Burg, mig.*

H. lacera (Michx.) R.Br.
La lacera v. terrae-novae Fern.
H. ~~Andrewsii White.~~
  Kil,Holy,Barred I.,JunctBr,MaryAL,Q,Doyle's,
PaB. *mig BB(fook)*

H. psycodes (L.) Sw.
  Freq. n.toNDBay and BB. Tab,Man,BB,Plac,GF,
BF,BO,Sum.*F9 mig. BB (Somerad) HB(RBK)*

H. fimbriata (Ait.) R.Br.
Gl,RP,Badg,Doyle's. *Lang.*

Pogonia ophioglossoides (L.) Ker.
  Freq. n. to NDBayand BStG. Carb,Don,WB,
BarredI,BC,Man,Whit,PaB. *LH^v Mig BB^v*

Calopogon pulchellus (Sw.)R.Br.
  Plac(Wil),OP,NHarb(Wag),Fogo,BarredI,
ShoalP,StG,Doyle's,PaB,Blom. *LH^v Lang BB*

Arethusa bulbosa L.
  Locally n. to NDBand BofI. Barred I,Q,
MJ,GF,Gl,MaryAL,LittleHarb,GrandL,PaB,
CStG. *LH^v, Burg, StB,*

Spiranthes Romanzoffiana Cham. *"Lily-of-the-Valley"*
  Com. TC,GF,Q,Holy,BC,FC,BarredI,Brig,Don.
Doet, Dophin, FS^v Lang BB^v

Epipactis repens (L.)Crantz (Antic.)

E. rep.,v.ophioides (Fern.)A.A.Eaton
  OpenHole,Aval(Osborn),Lot,NA,BofI,
MiddleA,Tab,StG,Steph,DanielsC(Trinity)
Doep^v Lang,

E. tesselata (Lodd.)A.A.Eaton,
  Whit,Salm,NHarb(Wag) Sum,SBF,BofI,Tab,
FlatBay(Bell), *Doet, Burd*

Listera cordata (L.) R.Br.
 Com. PaB,Salm,PS,GF,MaryAL,BadgBr,BenC,
Whit,Tab,Sum,TC,DR,BC,SBF,Brig,FC,Bum,
Doct,Lang,Pac,

L. auriculata Wiegand
 GF,BPS,Humb,HR,

L. borealis (Antic.)
 Eddy

L. convallarioides (SW.)Nutt.
 Freq.s. to NDB. Exploits and BStG,TC,
Tab,BenC,Marb,GF,BofI,FC,SC,Brig.Bear,Bum
Doct,Lang,Pac

Corallorhiza trifida Chatelain
 Freq. Sum,WoodsI,Tab,GF,MaryAL,StB,Doct
FC,Save Lang, med. Eddys Pac, Burt,
 C. striatorum Drej.
 FC,Big,Brand
C. maculata Raf.
 Salm,NHarb(Wag),Sum,BenoitC,CodRoy(Bell)
Doct,Lang,Eddys
 Malaxis brachypoda (Gray) Fern.
Microstylis monophyllos (L.)Lindl.
 Marb,GGard,FC,Bard,Doct, med, Eddys,Burt,
Bell

M. uniflora(Michx.
 Freq.n.toTrinityB,andBofI. BC,Sum,Carb,
Man,BitR. LHw Mig.

Calypso bulbosa (L.) Oakes.
 Doct. Eddys Burnt (aug)

SALICACEAE

Salix alba L.,v. caerulea (Sm.)Koch.
  StJ,BC.

S. serissima(Bailey)Fernald
  RP.

S. lucida Muhl.
  BF,SL,GP,GeoP,HR, *St.B, mig. BB*

S. luc., var. intonsa Fernald
  Gl,BF,Badg,BP,HR, *BB*

S. reticulata L.
  FC,SC,Brig,IB,Brig, *abund.on Str. MC, Cook, Sgg,*
Quirp, On PS Kal(aug)

S. vestita Pursh
  Croque and Hare Bay(LaPyl) PR, PS,GooseA,
Marb, Blom, Tab. *abund, on Str. Brand, Sgg. Quirp, On.*
Doet, MC, PB BB Kal(aug)

S. ves. var. erecta Anderss.
  Tab.

S. leiolepis Fernald
  Tab.

S. Uva-ursi Pursh
  Com. near w.coast. Look, Tab, PR, *Gund MC,*
Burin, Cook, Norm, Quirp, Lmt, GP, PaC, Doet BB(kus)

S. herbacea L.   (Gaspé)
  Doet.

S. anglorum, var. kophophylla Schneider.
   BB, Blom, No

S. angl., var. araioclada Schneider (Gaspé)

S. angl., v. antiplasta Schneider (Gaspé)

S. jejuna Fern.
   Big, norm, 4-Mi.
S. arctophylla Cockerell
   FC, Tab, Save, Brand, Sgs, Ice, Big, Cook, norm,
Boat, P&o

S. chlorolepis Fernald   (Gaspé)

S. gaspensis Schneider   (Gaspé)

S. brachycarpa Nutt.   (Gaspé)

S. cordif. var. encyda Fern
   FC, StB.
S. cordifolia Pursh
   IB, PR, FC, Doct.
S. cord. v. callicarpaea (Trautv.) Fern
   4-Mi, Doct, Quirp, FC, Big, Bear, Bard, Sgs, on ode.
S. cordifolia, v. Macounii (Rydb.) Schneider
   FC, Quirpon, IB, BP, Blom, Brig, SC, StB, GGard. No
Yank, Big, norm, Sav, Ice, Brand, Kal(an)s, P&o
S. cord. v. entonsa Fernald
   Doct, Yank
S. cordif. var. atra (Rydb.)
   FC, Brig, StB, GGard
S. cord. v. tonsa Fern.
   Ha, Doct.
S. cordata Muhl.

S. cord. v. Michauxiana Fern. ined.
   GF, HR, BF, SL, BPS, P&s

S. amoena Fern.
   Ha.
S. pedunculata Fern.
   Sav P&C.

S. myrtillifolia Anderss.
  IB? *Sav*, *Big*

S. myrt., v. brachypoda Fern.
  Tab? *On?* *Doet?*

S. latiuscula Anderss.
  Nfd,coll by LaPylaie.

S. glaucophylloides Fernald
  Humb?,HR?,Tab?,RomainBr,

S. pyrifolia Anderss.
  StJ,BBulls,Torb,Whit,GI,RP,C,BPS,GP,
?MistC,

S. obtusata Fernald   (Gaspé)

S. adenophylla Hook. (Lab.)

S. calcicola Fern.&Wieg.
  Pk,FC,SC,Brig.*SavC - Big, Conou Str, Burn, YM,*
SgS. Doet P?

S. laurentiana Fern.   (Gaspé)

S. candida Flugge
  com. near W.coast, BC,Blom,FrenchmC,CH,PS,
IB?,FC,SavC, *YM, Quirp, No?, B?, PuC?*

S. n. sp. barrens, JgB.

S. n sp.
PaC

S. cand., var. denudata Anderss.
   IB, CH, Ed

S. cryptodonta Fern.
   BPS, SavC, Burn.

S. Bebbiana Sargemt
   Com. n. toNDBay and BB. GF, Humb, BPS, BofI,
   MidC, LitRedIndL, BBulls, Sav, Hur BC

S. Bebb., var. perrostrata (Rydb.)Schneider
   Big, Ed, Sav, Had   (Mingan)

S. Bebb, v. capreifolia Fernald
   StJ, WB, Fogo, FC, BB, PaP. Gov, mid BB

S. Bebb,,v.luxurians Fernald (Gaspé)
   BerG

S. Bebb.,v.projecta (Fern.)Schneider
   WC.

S. pedicellaris Pursh
   Lmy, miq.

S. pedic.,v. hypoglauca Fernald
   MaryABr, SL, GP.

S.pedic.,var.tenuescens Fern.

S. hebecarpa Fernald (Gaspé)
Green Bay (Mrs).

S. humilis Marsh.

S. hum.,v.keweenawensis Farwell
  Com.n. to E.Br.ofHumber and NDB. Glen,
GF,SL,GP,StJ,Q,MaryAL,Cl,Whit,

S. discolor Muhl.

S. disc.,var. Overi Ball
Com. McIversC,Q,BC,RedRocks,RP,GF,StJ,BF,
WoodsI,CH,Torb,Fogo,StB,BBulls, *Cart, Doct, BB*

S. petiolaris
BB
S. paraleuca Fernald   (Gaspé)

S. planifolia Pursh
  BJ,MistC,StB, *Big, Hä, Boat, Ed, Bear, FC, Norm,*
Ed, Doct, Ral, Quirp, ON,

S. pellita Anderss.
  HR,GF,Gl,BF, *VBC*

S.pel.forma psila Schneider
  Q,Gl,RP,GF,GP,St,BPS,Humb, *Doct, BB*

S. viminalis L.
Cult.& Naturalized. McIversC,NewHarb,
SoDildo,BC.

S. n. sp. Wood, Mc Iver's Cove (high)

S. argyrocarpa Anderss.
  Nfd. acc. to Anderss.

Populus tremuloides Michx. *Freq. n.toNDBay and BofI. RP,WC,Tops, StJ. Lang.*

P. tacamahacca Mill.
  BP, *Doct, Ed, Lang MB*

P. tac.,var. Michauxi(Dode)Farwell
  Freq.near w.coast n.toIngornB. PS, Romaine Br.HR,Humb. *Burin, On.*

P. candicans Ait.
  BC.

MYRICACEAE

Myrica Gale,L.
  Com. OP, BB,Baccalieu,Fogo,MJ,CF,StJ,Glen, BenG,Chan,Balena, *595, Bay, LMt StP,*

M. Gale,var.Subglabra(Chevalier)Fernald
  Tab.

M. carolinensis Mill.
  Nfd. acc. to LaPylaie.
  StP,

BETULACEAE

Corylus rostrata Ait. ✓ North Harb R (Awa)
  Cl SmithSd(Cormack),Gl,NA,BF,GP,Q,BPS,
Humb,BC,BenoitC, Blom,FlatBay(Bell),
Cod Roy (Bell), Bay d'Espoir(LaPyl.) FJ
Lang. BB

Var. BB Hunt

Betula lutea Michx. f.   Witch Hazel
  Whit, Villa Marie, CH(Reeks), BB(Reeks),
Grand L(Murray),HumberR(Murray),WC,BC,
BenoitC,PaP,HR,Steph,FlatBay(Bell),
Anguille Mts(Murray),Crabb's(Howley),
CodRoy(Howley),Bay d'Espoir (LaPyl.) Lang.

B. papyrifera Marsh.
  Com. Badg. Man,StJ,Wat,OP,BC,Doct,Sauc,Ha,
Quirp, Burg, Lang, BB

B. pap.forma occidentalis (Hook.)
  SL,StJ, Sau, Doct

B.pap.var.cordifolia(Regel)Fernald
  Com. FC,BJ,BB,WB,StG,BayBulls StP, BB

B.pap.var. minor
  BB,TC,Blom, BB Doct

B. borealis Spach.
  Glen,BP,Blom,Tab,

B. microphylla Bunge.
  Tab,Blom,BB,StBarbe,Brig.Sauc, Doct, Burn,
Quirp, Big, StB, No, Burg, BB,

80

B. Michauxii Spach.
  Freq. n. to Quirpon I.and BofI. DR,Glen,
OP,Blom,SB,TC,GF,GP,Ben,Badg,MJ,Q,Look,
Musgr,GrandL,BarredI,GaffTops,PaB, Ral, Quirp
Burg, Burnt I. PB.

B. pumila L.
  Com. except in s.e.,where local. Badg,
CoalR,Blom,ShoalB,Baccalieu,Glen,Q,Chimn,
GL,BB,BP,IB,PS,Tab,GP,TC,JuncBr,BF,MJ,PaB,
CH,FC,Brig,PortugalC, SavC & Mt, StP,
PaC

B. glandulosa Michx.
  Q,FC,Blom, Doct, Quirp, BB.

Almus crispa (Ait.)Pursh
  Cl,BarredI, Doct

A. crisp.,var. mollis Fernald.
  Com. StJ,Tab,Blom,Chan,Balena,OP,Baccalieu,
BF,GF,DR,Glen,Tops,LitRedIndL,BC,PaB, Quirp
StP, BB

A. incana (L.) Moench,v.glauca Ait.
  Com. on main isl.n. to NDB,Humber and
BofI. Steph,GF,LitRedIndL,Badg,Tab,Q,
Sum,GP, Lang, BC

Fagus sylvatica L. esp panel (Ama)

URTICACEAE
Ulmus glabra Huds.
  nat. bears Kennie's K (Ama)
Urtica dioica L.
  Freq. weed. FrenchmC,GF,TW,Torb,StJ,
BBulls, StP

U. gracilis Ait.
  N. Pen. and W. coast to LaPH,CH,StB,
Doct, Bard.

U. viridis Ryob.

On̆, H̆a, FC̆, Beăr C̆, St̆B, CH̆, Lar̆k(N), P̆ruch, B̆B
pref

Ur̆tica urens L.
 Freq. weed Tw̆, GF̆, B̆B, PaB̆, Ar̆g, Q̆V, St̆G, S̆t̆R̆ PaC.

Laportea canadensis (L.) Gaud.

Lan̆, ~~Humulus Lupulus murays P(ama)~~

SANTALACEAE

Comandra Richardsiana Fernald
 FC̆, Brig, Blom, *MC̆*, SaŏC̆, Big, Brand, Boat, 5gg,
Burn, Ic̆e, PaC̆

C. livida Richardson
 Com. S̆L, T̆C, Bof̆I, GF̆, H̆I, PS̆, FC̆, Bad̆g, Tor̆b, St̆J,
Mist̆C, Brig, *Burn, Doct̆, F9̆* Pac̆, BB̆

LORANTHACEAE

Arceuthobium pusillum Peck
 Bof̆I, Tab̆, *seen generally from Pa B & Roy S, medium*
B̆B̆

POLYGONACEAE

Oxyria digyna (L.) Hill
 Grois I (La Pyl.) *Doct̆, S̆C*

Rumex Patientia L.
 Com. near settlements n. to NDB. and Bof̆I.
Irishtow̆n, Tor̆b, T̆C, Gaultois, *Bar̆t*,

R. occidentalis Wats.
 FC̆, SC̆, 5gğ, On̆, SaŏC̆, Baird, Doct̆, PaC(Roy)

R. Brittanica L.
  Com. n. to NDB.and BB. Whit,Bisc,SEA,
WC; *Gov,*

R. crispus L.
  Freq. weed. GF,BC,Trep, *StP, BB,*

R. pallidus Bigel.
  CE,DR,BB,WC,Sum, *Bard,*

R. mexicanus Meison
  Miq, BB,

R. obtusifolius L.
StJ,BJ,TC,BC,Chan, *BB,*

R. obt., var. sylvestris (Lam.)Koch.
  Trep, BC;

R. Acetosa L.
  Weed n. to NDB. and Rofl. Trep,BBulls,
Torb,StJ,Tops,Plac,Glen,Baccalieu,TW,BC,
Bay du Nord,FrenchmC, *Quirp, FY, BB,*

R. Acetosella L.
  Com. BC,StJ,GGard,FC,FrenchmC,Chan,Trep,
Savd, Burg, StP, BB,

Koenigia islandica L.   (Lab.)

Polygonum viviparum L.
  Com. s. to NDB.and CapeRay. Tab,IB,CH,
Baccalieu,Doyle's,LitR,MistC,FC,Brig, *Boat,*
Mank,StP, Mid BB,

P. lapathifolium L.
  GF;

P. lap.,var. salicifolium Sibth.
  Weed, Tops,Whit,GF,

P. scabrum Moench.
  QV,StJ,Cl,Lpt,TC,GF,BC, *LH, BB*

P. natans (Michx.)Eaton,
  Whit,BF,RP,BPS, *StP,*
  P. nat.f. Hartwrightii (Gray) Stanf.
  StP,
P. Hydropiper L.
  Tops, StJ,GF,BC, *Burg, BB,*

P. Hydr.,var. projectum Stanf.
  StP,

P. Persicaria L.
  Com. n.to DB.and BofI. BC,StJ,GF,Torb,
  Trep,GGard, *Burg, StP, BB*

P. Rali Babington
  StG,Pap,SEA, *miq*

P. acadiense Fern.   (N.S.)

P. boreale (Lange) Small
  StAnth,SC,Barred I, Funk I, *Bard, Ral, On,*

P. aequale Lindm.
  Wo

P. Fowleri Robmorn.
FC,NA,DR,Brig. Kal

P. allocarpum Blake
Burg

P. aviculare L.
Com. n.tonDB and BofI. GF,Torb,FrenchmC,NA
QVL,Sum,PS, S+P,

P. avic.var.vegetum Ledeb.
BC,SC,StJ,

P. avic.,var.angustissimum Meisn.
Arg.

P. sagittatum L.
Torb,MidC,Ren,Don,Tops,Kil,Whit,Salm,
Bonad(Cormack) S+P,

P. Convolvulus L.
Com.Weed. Whit,BF,BC, Bard S+P. BB.

P. cilinode Michx.
GF, Salmonier (Janet Murray
(drawn by a.m.)

CHENOPODIACEAE

Chenopodium capitatum (L.) Wats. (Gaspé)

C. rubrum L.
  Nfd.(Morrison)

C. ferulatum Lunell
   Wo

C. album L.
  Com. N. toNDB and BofI. GF, BC, StJ, McIverSC,
BF, BBulls, GGard, CB

C. lanceolatum Muhl.
   Wo

Atriplex maritima Hallier (Magd.)

A. glabriuscula Edmonston
  Freq. on coast. BofI, StJ, StB, IB, Burgeo.
GGard, LitB, Bard, SaVC, Burg, Gov.   Mig.

A. patula L.
  BC, Wo, S+P, BB

A. pat., var. hastata (L.) Gray
  StJ, NA, Prep, BBulls, DR, Arg, GGard, Gov, Burg,
S+P.

A. part., var. littoralis (L.)Gray (N.S.)

Salicornia europaea L.

sal. eur., var. prostrata (Pall.)Fernald
Stg, arg, Burg

Suaeda maritima (L.) Dumont.

S. Richii Fernald
DR,Arg,StG(LaPyl.)

Salsola Kali L.
FrenchmC,StG,LitR. *FJ, Burg rig,*

S. Kali,var.caroliniana (Walt.)Mett.
FrenchmC, *Burg.*

CARYOPHYLLACEAE

Spergularia rubra (L.) J.&C.Presl
Arg,StJ,Ferryland(Wag),Kil,Carb,Whit,
SpruceBrook,PaB,

S. salina Jr CPresl.
rig.
S. canadensis (Pers.) G.Don
Freq.on coast.FrenchmC, Na,Kil,SnA,PS,
Steph,Arg,LitR.*Bard, Ral, SavC, LH, Burg rig*

Spergula arvensis L.
Com.weed. QV,BC,GF,Tops,TorsCove,BBulls,
StP,

Sagina procumbens L.
  Com. near coast,Brig,CH,TC,StJ,Humb,PS, *Burg.*
BC,DR,Fogo,Glen,BarredI. *On, SC, On, Doet,SavC,*
GrBruit,   *Stp,*

S. proc.var.compacta Lange.
  GF,

S. saginoides (L.)DallaTorre (Gaspé)

S. *nivalis Lindbl.*   (Lab.)

S. nodosa (L.) Fenzl
  Com. s.to HDB and BStG.
CH,FC,StG,PS,Blom,BB,Fogo,BarredI, *Bard, Sgg,*
F3, Lan,

S. nod.,var.pubescens Mert&Koch.
  WH,TC,

Arenaria groenlandica Retz.(Lab.& Gaspé)

A. dawsonensis Britton
  TC,FC,Maro,Tab,GGard *Big, Sav, SC, Hea Pic*

A. uliginosa Schleicher (Lab.)

rubella

A.(verna L.,v.pubescens (C.&S.) Fern.)
Freq. s. to NDB and BStG,FC,Blom,CH,
TW,LB,TC,BB,Tab,Englée,Chimn,Musgr,Brig.*Big*
Barn,com.onStr. Doct, SJ5, Ao, Hak PdC

rubella var.

PdC
A. marcescens Fernald
BBS, Bloms,Coalk, *Ao,*

A. sajanensis Willd. (Gaspé)

A. peploides L.
PaP,

A. pepl,var.diffusa
Quirpon,FC, *On, Big, FJ,*

A. pepl.var. Maxima Fernald
BayofI(Eames.),WildCove(s.ofBofI),

A. pepl.var. robusta Fernald
Litk, *Bard, Burg, Set Lang. PdC BiB,*

A. cylindrocarpa Fernald
Blom's,Coalk,Tab,BBs,Brig,*Sc, Saoc, Big, Burn*
Brand,Boat,YM. SJ5, On, Ice, Ao,

A. lateriflora L.
　Freq. CoalR,Salm,Salm,StG, IB,BofI,
FrenchmC,DR,CH,FC, Hā, Burg, Nuiŕ Pat

A. macrophylla Hook. (Lab.Gaspé)

Stellaria borealis Bigel.
　Com. DR,GP,BarredI,ShoalP,WC,CH,SC,
Arg,CoalR,FC,GGard, Brand, On, Cook, Doct,
Pic

S.bor.var.Bongardiana Fernald (Gaspé)
S. bor. var. floribunda Fern.
　　Doct Bap
S. uliginosa Murr.
　Com. n.to NDB and Codroy.　BJ,CGrb,Trep,
Torb,MidC,Plac,Doyle's,Kil, FC,Lpt, Sap

S. crassifolia Ehrh.
　Quirpon,FC,StB,Brig,PS, Big, Bard, Kal, SC,
Pac

S. humifusa Rottb.
　DR,FC,StB,Brig,PS,StG,steph, Boat, Bard, Kal,
LH, Burg, Pac

S. longipes Goldie
　Com. on W. coast s.to IngornB, FC,PS,
IB,StB, On Pac
S. longip. v. Laeta (Richardson) Wats
S. longip. v. Edwardsii (R.Br.) Wats.
S. florida Fischer　(Gaspe)
　On

90

S. longifolia Muhl.
    Salm, *Lang.*

S. graminea L.
    Com. D. to NDB and BofI. FrenchmC, Gl.
StJ, BC, GF, BBulls, *Gov, StP. BB.*

S. media (L.) Vill.
    Com. Trep, Funk, BarredI, Glen, FrenchmC,
GF, BC, CoalR, StJ, Chan, *Burg, StP.*

Cerastium alpinum L.   (Lab.)
    On.
C. *aep., var. glanduliferum Koch*
    Burn.
C. alp., v. lanatum (Lam.) Hegetschw.
    Quirpon, StAnth. *Boat, Burn, Ha, Quirp Big 4M*
    Pal (aug)
C. *Regelii Osterp.*
    Brand
C. Beeringianum C. & S
    IB, FC, PS, GGard. *Sc, Sgg, Doct, Norm, Boat, Big*
Green.

C. Fischerianum Seringe   (Gaspé)
    Big, Pal.

C. terrae-novae Fern. & Wieg.
    Blom s, BBs, CoalR, *No.*

C. vulgatum L.
    Com. Fogo, GrandL, BC, CoalR, MJ, BF, StJ, FC,
StP. Ha

B. ?
Pal.   *10*

C. viscosum L.
Burgeo. *8th.*

C. arvense L.
Com. s. to NDB and BStG. Blom, BBs, DR, Barred I.
PC, Snooks, TC, WH, Brig, *FC, Mc, Doct No bushes*

C. cerastioides (L.)   (Gaspé)
Doct.

Agrostemma Githago L.
NewH (Wag.)

Lychnis alpina L. *Sweet William*
PA, WH, TC, BBs, Blom s, CoalR, PipestoneP(Howley)
No

L. apetala L.   (Lab.)

A. affinis Vahl   (Lab.)
L. dioica
   introduced (Anna)

Silene acaulis L., v. exscapa (All.)DC.
Com. s. to NDB, Bay d l'Est R. and B of I.
Baccalieu, WH, TW, Quirpon, SC, FC, PR, BBs, Blom,
PipestoneP(Howley). *Doct, Quirp, Dog Pen.*
FS. No, Mig, Mid, BBs
S. ac. v. subcaulescens

S. noctiflora L.
BF, BC, Spruce Br.

92.

S. Armeria L.
  Deer Lake (Wag.)

S. latifolia (Mill.)Britt.&Rendle.
  GF.

PORTULACACEAE

Montia lamprosperma Cham.
  Whit, BarredI Quirpon, FC, Brig. *Boat, Ed, Bard*
  Doct, Burg, StP, Pt'c

M. rivularis C.C.Gmel.
  HolyI.

Claytonia caroliniana Michx.
  ShoalPoint(near Chimn),GrCodRoy(Bell),

C. carol.,v.sessilifolis Torr.(Gaspé)

NYMPHAEACEAE *Pond Poppy, Beaver More,*
                                    Beaver Root,
Nymphozanthus variegatus (Engelm.)Fern.
  Com. n. to Croque and BofI.FrenchmC, PaP,
  Blom, BF, OF, Whit, GF, BC; *Ral, Pond Doddy*
  Burg, StP, Boat, GF

Nymphaea odorata Ait,
  Com. n. to E.BrHumb,Exploits andCroque,
  Torb.

N. odor.,var. rosea Pursh
  Com. n. to Exploits and BStG. BJ,Whit,GF,

Nymphoz. rubrodiscus (Morong) Fern-
     Prince's LookOut (Ama)

RANUNCULACEAE

Ranunculus aquatilis (Complex, to be coll.)

R. hederaceus L.
   QVL, MidCove, StJ, Carb, NewHarb, OpenHole(Osborn)

R. Cymbalaria L.
   Com. on coast. FC, PaB, DR, Steph, CH, IB, Chan,
Chimn. *Mig*

R. Pallasii Schlecht.   (Lab.)

R. Purshii Richardson   (Gaspé)
   R. Gmelini Ca. Meyer
   Rob. Br (RBK)
R. hyperboreus Rottb.
   FC, SavC, *SC, Bard,*

R. lapponicus L.   (Lab.)

R. reptans L.
   Com. GF, FrenchmC, WB, StJ, Salm, GrandL, Over-
falls, MaryABr, GI, RP, IB, Tab, BarredI, Trep, *Bard,*
S&P,

R. rept. var. ovalis(Bigel.)T.&G.
   StJ, *SC, SGg, Lary,*

R. Flammula L. *Lang*
  StJ(Wag),QVL.

R. Flam.,var.angustifolius Wallr.
  QVL.

R. nivalis L.   (Lab.)

R. Allenii Rob.   (Lab. Gaspé)

R. pygmaeus Wahlenb.   (Lab. Gaspé)

R. pyg.,var. petiolulatus Fern. (Gaspé)

R. pedatifidus J.E.Sm.,v.leiocarpus(Trautt.)
  Ha, On, moist rocky backs, Ral(Aug) Fern. (Lab.)

R. abortivus L.
  Freq. FC,IB,FrenchmC,BC,GF,ShoalPt. *On Doct*
  Bb

R. abort., var. eucyclus Fern.
  StG. *Doct,*

R. sceleratus L.
  HarbGrace(Wag),NewHarb(Wag),SmithSd(Cormack),BarredI,Lpt. *Lang,*

R. recurvatus Poir.
  HumbR(Wag),PaP.

95

R. pennsylvanicus L. f. *Bard*
  NA,BF,CodRoy(Bell), *Bard*,

R. Macounii Britton,
  BF,GF,DeerL,HawkeHarb,Chimn,WC,Meadows(Wag)
  Doct, City

R. septentrionalis Poir.   (Gaspe)

R. repens L. *Gillcaps*
  Com. Man,IB,BC,GF,

R. rep.,var.glabratus DC.
  Gl, ~~Harmp Bx (RSt)~~

R. rep.,var.villosus Lamotte.
  Torb,Chan,BC,Salm, *Burg, StP, Ral(AmJ)*

R. rep.,var. erectus DC.
  Tops.

R. acris L. *Gillcaps*
  Com. Tops,StAnth,FrenchmC,GF,GrandL,BC,
  StJ,BB,Gl,StG, *Fc, Cook, StP, Lang,*

Thalictrum alpinum L.
  Com. s. to ExploitsR and BStG. FC,Badg,
  Q,Blom,MJ,HR,IB,Musgr,Brig, *Burn, Quirp.*
  Ral

T. confine Fern.   (Lab.)

T. *dioicum L.*
  StP,   *False Maidenhair* *Columbine*
T. polygamum Muhl.Complex to be watched.
  Quirp. Doct, LMo, FJ, Burg, StP, Mig

Anemone parviflora Michx.
  Com. s. to ExploitsR and BStG. Tab, Marb,
GF, PS, FC, SC, Brig, Blom, Ed, Doct, Burin, No'. *Han*
Mid BW

A. multifida Poir, var. hudsoniana DC.
  Snooks, GF, *Doct, Han (dominant) BB*

A. riparia Fernald   (Gaspé)
  HR (RBK)

A. Richardsoni Hook.   (Lab.)

A. canadensis L.   (Gaspé)

Clematis virginiana L.   (Gaspé)

C. verticillaris DC.   (Gaspé)

Caltha palustris L. ✓ *Spring Cowslip*
    Com. on w. coast. FC, McIverSC, SC, StG,
  BB, *E. to Bonk, Doct, In Gardens at Burg, & Pa B.*

Coptis *groenlandica (Aeau) Fem.* Snakeroot
  trifolia (in) Salisb.
    Com. TC, Balena, BC, Whit, Blom, Brig, FC, *Quirp*
SC, Doct

Aconitum Napellus L., v. Lobelianum Reichenb.
  Brig, StJ, *upper Gullies (Ana)*

A. Nap. v. Lob, f. albiflorum Reichenb.
  Brig,

Aquilegia vulgaris L.
  Gr. Jo Beachy Cove (Ana)

Actaea rubra (Ait.) Willd.
  Com. TC, Badg., Coalr, PR, Balena, SB, FC, Gl,
ShoalPt, *Brig.Big. Brand, Quirp. Dock, BB*

A. rube var
*   Pac*

BERBERIDACEAE

Berberis vulgaris L.
  Nfd. acc. to Hook.

PAPAVERACEAE

Papaver radicatum Rottb.   (Lab.)

P. somniferum
     Bard,

FUMARIACEAE

Dicentra Cucullaria (L.) Bernh. (Gaspé)

Corydalis sempervirens (L.)Pers.
  BF, GF, *Buchan (KPg)*

C. aurea Willd.   (Gaspé)

Fumaria officinalis L.
  StJ, NewHarb(Wag), *StP, Topsail (E.Holloway),*

CRUCIFERAE

Sisymbrium altissimum L.
  StJ.

S. officinale (L.) Scop.
  StP,

S. off.,var. leiocarpum DC.

Descurainia intermedia (Rydb,)Daniels
                              (Gaspe')

D. Hartwegiana (Wats.) Britt. (Gaspe')

D. Sophia (L.) Webb.   (Gaspe')

Subularia aquatica L.
   Whit, Plac, GF, RP, Badg, BC, Briqu(Ang)
   OwnsVidi (nma)

Lepidium sativum L.

L. virginicum L.
   GF,

L. apetalum Willd.
   Trep,

L. ruderale L.

Coronopus didymus (L.) Sm.
   St.J, HarbGrace,Carb,Whit,

Thalspi arvense L.
   StJ, Carb, Cowche(Wag.),BBulls, On, mig,

C — ?
535

Cochlearia cyclocarpa Blake
    StJ(Wag) Funk,Baccalieu,Barred I,Fogo,
BlackI, WH,CH,StG, *Wo, Watts's* *StP, Brig. BB*

C. tridactylites Banks
    DR,FC,PS,StB,Brig, *Pat*

C.
  beaten

C. fenestrata R.Br.   (Lab.)

C.
  Bard,

C. groenlandica L.   (Lab.)
  Ral (Aug).

Eutrema Edwardsii *R.Br.* (Lab.)

Cakile edentula (Bigel.) Hook.
    IB,CH,CoalR,StG,FlatBay(Bell),Brunette I(Wag)
FrenchmC,LitR, *Bard, FB, St, Brig. StP, BB*

Brassica campestris L.
    StJ,GrandL,

B. nigra (L.) Koch
    StJ(Wag), *Burg, StP,*

B. juncea (L.) Coss.
    Bard,

B. arvensis (L.) Ktze.
    GF,Don,GCard, *StP,*

B. arv.,var. Schkuhriana (Reichenb.)
BC.

2 B. repanda (Willd.) DC. !
   " Hard. du Croc", Pl 9-10 cm. high; fl. apet; las real, 3 cm. long, dentoled; fr. lineal, 2 cm. long.

Diplotaxis muralis (L.) DC.
Coll by Jensen

Raphanus Raphanistrum L.
   QV, StJ, Carb. S+P,

R. sativus L.

B. arbarea vulgaris R.Br.
   Records by Wag. S+P,

B. vul., var. longisiliquosa Carion
   StJ.

B. vulg.,var. brachycarpa Rouy & Foucaud
B. verna (Mill.) Aschers,
   S+P,
B. orthoceras Ledeb.
   BF, GF, Sav C, Quirp.

Ropira Nasturtium-aquaticum(L.)Schinz &
                           Thell.
   QV, StJ,

R. Armoracia (L.) Hitch.
   S+J (Aura)

R. sylvestris (L.) Bess.
  Ren, StJ, Geob,

R. *hispida (Desv.), var. glabrata Lunell*
  ~~palustris (L.) Bess.~~
  MaryAL,

R. ~~pal. var. hispida~~ *hispida* (Desv.) *Britton*
  BF, GF, RP, Badg, DeerL,

Cardamine bellidifolia L. Gaspé)

C. hyperborea Schulz   (Arct.)

C. pratensis L.

C. prat., var. angustifolia Hook.   (Lab.)
  7C, SauC, Deadm, ShoalC, Cook, Braud, Boat, Ed,

C. prat., var. palustris Wimm. & Grab. *Mayflower,*
  GVL, Ren, StJ, *Doct, Brf g, Tops (ama)*

C. hirsuta)
  ~~Hirsuta Hout~~

C. pennsylvanica Mghl.
  Com, GF, StJ, Whit, RP, Marb, *Doct, Lauz, BB*

Lesquerella arctica (Wormskj,) Wats.
  PR, Tab, GreenC, *Big, Burn, Cook, Boat, 5g9,*
  Pac W

Hymenolobus procumbens (L.) Nutt.
Quirpon, Conche, Cook.

Pick-pocket

Capsella Bursa-pastoris (L.) Medic.
Com, Gl, Chan, BC, Fogo, StJ, SC, FrenchMC,
Trep, Burg. StP,

Camelina sativa L.

C. microcarpa Andrz.
StJ,

Neslia paniculata (L.) Desv.
StJ,

Draba stenoloba Ledeb.   (Lab.)

D. crassifolia Graham   (Lab.)

D. arctica Vahl.   Gaspé)

D. alpina L.   (Lab.)
Nfd. acc. to LaPyl.
D. oligosperma Hook (Gasp.)

D. fladnizensis Wulfen (Gaspé)

D. nivalis Lilj.   (Gaspé)
Ha, On, Quirp, BB,

D. aurea Vahl   (*Lab.*)

D. *luteola Greene* (*Bic, Mingan*)

⌐ D. incana L.
    BarreďI, Exploits, PŠ, GreeňG, Nořm; *Pic*

⌐ D. inc.,var. confusa (Ehrh.)Poir.
    FČ, Brig,PS,CH,IB, *Baŕd, Sjȷ̌, Bočt, Buřn, SČ, Baŕd,*
Biȷ̌

⌐ D. stylaris JGay
    Chiḿ,

⌐ D. megasperma Fern.&Knowlt.
    PǍ,Stǰ,Brig, *FČ, Savě, Biȷ̌, Boǎt, Nořm, Braňd,*
Greěn, Quiřp, Pic Graš,

⌐ D. arabisans Michx.
    FČ,Snooks,GF,Musgr,John'sBeach, *Wočt,*
medřus Hai Mil BB

⌐ D. arab., var. orthocarpa Fern.&Knowlt.
    RopeCove,

⌐ D. pycnosperma Fern.&Knowlt.(Gaspé)
    Dočt,

⌐ D. hirta L.
    DŘ,PA,StAňth,GGard, *Dočt, Sjȷ̌, Oň, Hǎ, Buřn, Boǎt,*
Quiřp, Braňd, Mauvě

⌐ D. rupestris R.Br.
    TW̌,TǑ,PŘ,SČ,FC,Brig, *Biȷ̌, Buřn, Dočt, Baŕd, Hǎ,*
Quiřp, Ice, Deǎd, Savě, Fǧ

Arabis humifusa (J.Vahl) Wats.   (*Lab.*)

A. hirsuta (L.) Scop   (Gaspé)

104

A. alpina L.
 Croque (LaPyl),StAnth,FC,PR,CH, Sav. Big.
Doct, Quirp, On- Anch. Pac Gas.

A. Collinsii Fernald   (Gaspé)

A. Holboellii Hornem.   (Gaspé)

A. Drummondi Gray
 GF, Doct BB

A. Drummondi v.connexa(Greene)Fern.(Lab.)
 On, Doct
                   coarctatum Fern
Erisymum asperum (Nutt.) DC.
 Chim

E. cheiranthoides L.
 StJ, Carb,Trinity(Wag),NA,BF,GF,PR,
BC,

Hesperis Pallasii (Pursh) Fern. (Arct.)

Hesperis matronalis L.
 StJ

Parrya arctica R.Br.   (Arct.)

             Longii Fern.
Braya purpurascens (R.Br.) Bunge
 SC, Sav. Yank
 B. Richardsonii (Rydb.) Fern.
 Boat
B. humilis (C.A.Meyer)Rob.
 Tab
 B. americana (Hook.) Fern.
 Sav.Big. Burn. Norm, Boat, Ice, 4M, Watts,

Conringia <u>orientalis</u> (L.) Dumort.

<u>SARRACEMACEAE</u> *Indian Jug*

Sarracenia purpurea L. *Indian Cup.* *Ind. Pipe*
Com, Frenchm C, OP, Blom, Chan, MJ, Sum, GF,
Badg, Balena, BB, Whit, PaB, Trep, FC, *Ral, LH'.*
to pubesp. LMt. S+P.

f. heterophylla
BB

<u>DROSERACEAE</u> *Muskrat Flower*
Flycatcher   Larkspur
Drosera rotundifolia L.
Com, Steph, BB, Blom, OP, FC, CV, Tops, PaB, StB,
Barred I, BJ, Trep. *Quirp, Burg, S+P.*

D. anglica Huds. ✓
Croque(Banks), BF, MaryAL, MJ, C, FC, Brig, PR,
Look, Blom, *Big, Quirp, OtkiP, Bart, Pal BB.*

D. obovata
BR

intermedia Hayne
D. ~~longifolia L~~
Com, n, to WhiteB and BB. Barred I, Blom,
OP, GF, RP, BF, GI, MJ, Man, Whit, Badg, Look, BC,
S+P.

D. linearis Goldie
Blom. *Pac*

<u>CRASSULACEAE</u>
Tillaea aquatica L.
Arg.

Sedum acre L.
  StJ(Wag)

S. Telephium L.
  StJ(Wag),BC.

S. stoloniferum Gmel.
  StJ.

S. roseum (L.) Scop.   *Midsummer Men, Houseleek*
  Freq. on coast, TC, WH, Tw, BarredI, Fogo,   *Scurvy Grass*
  CH, FC; *Fg, Miq. Mia. Pat. BA.*

S. villosum L.   (Lab.)

                SAXIFRAGACEAE

Saxifraga rivularis L.
  StAnth(Wag), *Doct, Quirp.*

S. cernua L.   (Gaspe')
  Nfd. acc. to Pursh

S. cespitosa L.
  Englee, FC, Brig, PS, CH, *com on St. Boat, BarI, On,*
  Burn, Cook, Brand, Doct Pat BA.

S. stellaris L.   (Lab.)

S. stel. var.comosa Poir (Me,Lab.)
  Doct.

S. hieracifolia Waldst. & Kit.(Lab.)

S. nivalis L.   (Lab.)

S. gaspensis Fern.   *(Gaspé)*

S. aizoides L.
   Com. s. to Canada Bay and God Roy. Blom,
Marb, BC, FC, BB, IB, SC, Brig, RopeC, *SJ, On,
Doct, FJ, Mid, Pic Brat B0(dint) remaining main R/18.*

S. tricuspidata Rottb.   (Lab.)

S. Hirculus L.   (Lab.)

S. flagellaris Willd.   (Arct.)

S. Geum L.
   Nfd,(Steinhaur in herb Paris)

S. Aizoon *Jacq.*
   Com. s. to NDB.and Red Rocks, TC, PA, GF,
SB, CH, Chimn, GGard, *Sav, Braud, Doct, On,
Hah mid Pic BB(dint).*

S. oppositifolia L.
   Com.s.toCanadaBay and BStG. Tib, Herb, SC,
BB, IB, FC, Brig, GGard, *Brat, Doct, FJ,
Ant, mid Pic B0(Cinit)*

Mitella nuda L.
Com. Ben.C.RedRocks, TC,Gl,FC, Badg,
BC, Frenchmc, *Doct Lang, Pac*

Chrysosplenium americanum Schwein.(Gaspé)

C. alterniflorum L.,v. tetrandum Lund.
(Arct.)
Parn. caroliniana
Eddy Brit. BR
Parnassia Kotzebuei C.&S. (Lab. Gaspé)
IB? (LaPyl.) *On, Doct*

P. parviflora DC.
StB,Brig,IB,CH,Chim,GooseA,Bofi,PaP,*Doct*
SJs.F9 Mid Pac Eddy BR

P. ~~palustris L.~~ *multiseta (Ledeb.) Fern.*
Quirpon (LaPyl.),FC, IB(LaPyl), *S+B, Ral,*
Otto P.Cook

Ribes hirtellum Michx.
Com. RP,DR,GF,BC,Fogo,BarredI,StG,
Tab,Tops, StB,Brig,*FC, Doct, StP, BB Ral(Amf)*

R. hirt.,var. saxpsum (Hook.) Fern.
Crabb's, *Pac,*

R. nigrum L.
Don,Carb, *StB.*

R. lacustre (Pers.) Poir
    Com. FrenchmC,GF,LarkH,T b,BofI,BC,StB,FC,
Avet

R. ~~prostratum~~ (He). *glandulosum Grauer*
    Com. StJ,GF,WB,Fogo,Bccalieu, BofI,BC,
BS,Chan, BB,CoakH,HopeCove, *StP, Miq. GB*

R. rubrum L.
    Esc. from cult. HarbGrace (Wil.) *StP*

R. triste Pall,
    FrenchmC, Brig,FC,

R. trist.,var. albinervium(Michx.)Fern,
    Com. near W. coast e. to Br.Humb. GF,
FrenchmC, BB,BC,Crabbs,

                    ROSACEAE
                              Deadman's Flowers.
Spiraea latifolia (Ait.) Borkh.
    StJ,

S. lat.,v. septentrionales Fernald
    GF,StJ,Tops,Carb,BP, *Burg StP,*

Sorbus americana Marsh. *Dogberry Catberry Eightberry*
    Com. GF, FrenchmC, BB, MJ,Cl,StJ,Tops,CH,
OHellP, BD,

    S.
Dogb. H.

110

S. dumosa Greene *Dogberry*
    Com. Gl, MJ, BC, StJ, ., GP, BC, ShoalB,
    Tops, Whit, BC, Chan, MohnsBeach. On, S+B.
    S+P,

× S. *Arsenii Brit.*
    mig.

S. ~~ordatifolia, the.....~~ *floribunda (Black)* *Wild Pear, Winter Pear*
    Com. n. to N+B and BofI. StJ, MJ, RP, GP,
    Q, DR, RP, Blom, StG, GP, BP, Steph, GP, Fogo,
    QV, BarredI, Bracelieu, *LH* *S+P.*

S. melanocarpa (Michx.)
    Badg, StG, (Eames), FrenchmC.

Malus -
    StG (Ames)

Amelanchier sanguinea (Pursh)DC. (Gaspé)

A. Fernaldii Wieg.
    Tab, GreenC, *S+* *R&C*

A. stolonifera Wieg.
    GP, RP, Trep,

A. laevis Wiegand *Indian Pear*
    StJ, Whit, GP, FrenchmC, Don, GGard, *79* *S+P.*
    Hal M)

A. laevis, v. nitida (Wieg.)Fernald.
    StJ, Gl, BP, GP,

A. Bartramiana (Tausch) Roem.
    Com. Snooks, StJ, Don, TC, BP, GP, HH, MJ,
    LRed IndL, Whit, Carb, Gl, DR, RP, C, GaffTops,
    BC, FrenchmC, *Aaron, Quirp, Burg, LMt, S+P, Hab*
    &C

gus laurentiana Sarg.
BF,RP, Hughs,

C. macrosperma Ashe
BF,RP,

Fragaria virginiana Duchesne
Freq. n. to NDB. and BofI. Torb,
Don,Gl,GF,MJ,BC,BBulls, Burg,

F. virg. var. terrae-novae(Rydb.)Fern.&Wieg.
Com. Man,Holyr,FrenchmC,Don,GF,IB,CH,
BC,LarkH,SavC,,CHim,StB, FC,Busn, OK. Mist
StB,

F. vesca L.
Don,Tops,Kil,whit,Cl,NA,GF,BC,HR,

F.vesca,var. americana Porter (Gaspé)

Sibbaldia procumbens L. (Gaspé)
Doat,

Potentilla fruticosa L. Willd,
Com. GF,TC,G,MJ,WB,HB,BB,Gl,Cerb,PA,
Musgr,CH,BarredI,StG,Man,Salm,GGard,FC, On
L Mt, Miff,

P. palustris (L.) Scop.
Com. PS,RP, Baid, On. St. StP,

P. pal.,v. parvifolia (Raf.)Fern&Long.
Cuirpan, BB,FC, Diet,

P. pal.,var. villosa (Pers.) Lehm.

P. tridentata Ait. ✓
   Dirt com. GP,BarredI,Gl,OP,FC,BB,
Blom,Carb,BP,TC,Fogo,Tops,Musgr,StJ,
Trep, *Brand. Belg. StP*

P. pectinata Raf.
   Chimn. *Doct. Wb*

P. nivea L.
   PR, Tab,GreenG, *Doct. Norm. mid BB*
P. niv. var. macrophylla Ser.
   Tab. Doct. Sogg. Gr. Ha Burn Cook, Sch, Pac
P. pulchella R,Br. (Arct.)
P. ustigapensis Fern.
   Burn
P. rubricaulis Lehm. (Arct.)

P. Vahliana Lehm. (Arct.)

P. emarginata Pirsh (Arct.)

P. argentea L.
   Brigus (J. Knowling)

P. recta L.

P. norvegica L.

P. norveg.v.hirsuta (Michx.)Lehm.
   Com. FC,TC,GF,PS,Gl,Hawke,Humb,Fogo,
BBulls, Brig, Gault.*Aoch.SC.OK*   *Lang, BB*

P. norveg.,var.labradorica (Lehm.)*Fern.*
   Big.Dock.BB

P. fragiformis Willd. (Gaspé)

P. Ranunculus Lange (Lab.)

P. Robbinsiana Oakes (N.H.)

P. alpestris Hall. *fr*
   DR. GreenC.*FC,SaoC,Big, Burn,Cook,Braud,Norm.
Boat,Dich,On,Sgg. Plac grassy meadow,Kal(Aug)*

P. procumbens Sibth.
   StJ,Ferryland(Wag.),BBulls,
   P. erecta/P. Haufi
   (Vtime)
P. canadensis,v.simplex(Michx.)T.&G.
   RP.

P. Anserina L.
   DR.,BF,Pa3,FC,*Ed SrP.*

P. Ans.,var. sericea Hayne
   Bor
P. Egedii Wormsk. (Lab.)

P. pacifica Howell
   Com. on coast. N.A.,StG,IB,FC,FrenchmC,
Kil,CH, Plac, Chan,BarredI,

P. sterilis (L.)Garcke
Cliffs, Brigus(AMA. murrays P(Kind)
                    (painting)

114

Filipendula Ulmaria (L.) Max.   *Princes Feather*
   BC, *Beachy Cove (Ama)*
 F. hexapetres Gilib.
   er (Ama).
Geum macrophyllum Willd. *Blood Root. Jack Root*
   Com. Don, PA, Snooks, BC, ShoalP, Chimn,
FC, SavC, *Doct, Quirp, Lang, Pac, Ral (Amg)*

G. rivale L. *Chocolate Root*
   Com. GF, MJ, FrenchmC, FC, Gl, *Miq, Pac, Ral(Amg)*

? G. triflorum Pursh
   Nfd. acc. to Hooker

Dryas integrifolia Vahl.
   Com. s. to Canada Bay and BStG. Englée,
SC, FC, Brig, PR, PS, Tab, GreenG, *Doct. Quirp. Ham*
Miq Pac. 1913

D. integr., var. canescens Simmons
   Tab, GreenG, *Pac,*

D. Drummondii Richardson   (Gaspé)

Rubus idaeus L.
   StJ, *Doct*

R. id., var. strigosus (Michx.)Max.
   Com. BB, WB, Whit, Torb, GrandL, Glen, DR,
Baccalieu, StB, *Quirp*

R. id., var. canadensis Richardson
   Com. GF, Chan, BC, StB, *StP,*

R. Chamaemorus L. *Bake Apple*
  Com. DK,MJ,OP,Chan,Whit,BarredI,Fogo,
Salm,BaLena,PaB,FC,CStG,Brig,BB, *StP, Dnt.*

R. arcticus L. *Plumboy*
    FC,GF,Gl,Brig, *Sasc. Big Burn. Com. on StP,BB.*
SgB Onig Pal Bat

R. acaulis Michx. *Plumboy*
    Com. s. to Gander R., Exploits R and
Cow Head. GF,RP,IB,FC,StB,PS, *Dock Quirp Miat*
StP, Pal Brist

R. pubescens Raf. *Plumboy Dewberry, Swampberry*
    Com. GF,MJ,FrenchmC,Gl,Gl Badg, Baccalieu,
WB,BC,BB,Man,FC, *Ha. Madie, Burn, StP,*

R. canadensis var.septemfoliolatus Blanchard
    Torb,StJ,KaneVal,Tops,Kil,Gl,NA,BF,GF,
BPS,GF,

R. recurvicaulis Blanchard *English Blackberry*
    StJ,BJ,Whit,NHarb,OF,Trep,PaB, *StP,*

Alchemilla vulgaris L.

A. vulgar.,v.filicaulis(Buser)Fern.&Wieg.
    Com.on w.coast and to Quirpon. FC,ShoalPt,
SB,Musgr,RomaineBr,StG,Doyle's, *Dock, Quirp*
Hunt. Dock, BB meadows, Ral(Aug)

A. vulg..v.vestita (Buser)Fern.&Wieg.
    FC, *Dock.*

A.vulg.,v.grandis Blytt   (Lab.)

A. alpina L.
(~~Old report from~~ Miquelon)

Agrimonia striata Michx.
Salm,Chim,YorkH (E.&G.) *LH*

Sanguisorba canadensis L. *Tobacco Leaf Indian Tobacco*
Com. Chan,PS,GF,TC,BC,OP,Look,Blom,FC,
BB,Tops,Badg,Man,PaB,Whit,FrenchmC; *79*
StP,

S. can.,var.latifolia Hook. (Antic.)

Rosa nitida Willd. *Dog Rose*
Com. n. to NDB and BofI. Blom,Don,Torb,
Chan,Gl,GF,G,KB,Fogo,ShoalB,BarredI,StJ,PaB,
Trep, *SJ StP,*

R. carolina L.
StJ,Tops,Kil,BJ,Carp,Gl,DR,BF,GF,
ShoalP,StG, *Bur StP,*

R. virginiana Mill.
Com. n. to Bonavista Bay and BonneB.
StJ,Kil,HarbGrace(Wil),Carb,Cl:OpenHole
(Osborn),SEA,Sum,WC,Steph,StG,

R. Eglantina
Curl.

Prunus virginiana L. *Chuckley Plum. Mazzard*
Com. n. to NDB.and BofI. G,RP,HR,Musgr,
Salm,Tops,LitRedIndL, *L Nfd Lang,*

B. pensylvanica L. f.
Com. n. to NDB and BB, StJ,LarkH,WB,StG,
Fogo,Tops,Whit. *79 StP.*

LEGUMINOSAE

Lupinus arcticus Wats.   (Arct.)

L. nootkatensis Donn
    Cl, *murrays P(ama)*

Trifolium pratense L.
    Com. FrenchmC, GF, GI, CH, BC, GrandL, Whit,
Baccalieu, *StP*

T. prat., v. frigidum Gaudin.
    Lff

T. repens L.
    Com. FrenchmC, GI, GF, BC, GrandL, Whit, StB, *StP*

T. hybridum L.
    Com. BB, FrenchmC, GF, GrandL. Whit, CoalR,
miq,

T. agrarium L.
    BF, GF, HR, *LH*

Melilotus officinalis (L.) Lam.
    Heart's Content (Wag.) *LH*

Medicago lupulina L.
    StJ, GF,

Lutsoynie

Astragalus alpinus L.
    Chim, *Burn Quirp*

a. stragulus Fern.
    Cook

A. alp., var. Brunetianus Fern.
  BF, GF,

A. frigidus (Richardson) Gray,
  var. americanus (Hook.) Wats.   (Gaspé)

a. ?
  mid

A. eucosmus Robinson
  BF, GF, Bagg(Wag.) *Burn, Ha, Cook, Braud, Boat, Quin.*
  Ori, Mauve, grassy meadow near beach, Raf(aug)

a. Euc., var. facinorum Fern.
  GFri
A. aboriginum Richardson   (Gaspé)

a. Blakei Eggl.
  Doct
Oxytropis ~~campestris (L.) DC.~~ *Maydelliana Trautv* (Lab.)

terrae-novae Fern.
O. ~~anonyma coerulea Hook.~~
  Freq. s. to WhiteBay and BStG. SC, IB,
Chimn, Tab, GreenG, *SavC. Bif, Burn, Cook, Norm, SJd, Quin,*
Pat, Two,
O. ~~anny var~~ *johannensis Fern.*
  CH, Ori, Wa, Doct, Green G., BB tooke, gravelly banks, Raf(aug)
? O. arctica R. Br.
  CH, ~~Greens~~,

O. Bellii (Britton) Palib.   (Lab.)

O. arctobia Bunge   (Arct.)

O. podocarpa Gray   (Lab.)

O. foliolosa Hook.
  Burn, Cook, Braud, mid
Hedysarum alpinum L.
  SC, CH, Tab, GreenG, *SavC. Bif, Burn, Cook, Norm, YM,*
Quin, Ori, Pic, Bm

H. alp., var. americanum Michx.
  BF, GF, WC, *Doct. mid.*

H. Mackenzii Richardson
  GreenG

Vicia angustifolia Reichard
  NHarb (Wag.)

V. ang., var. segetalis (Thuill.) Koch
  GeoP. *S+P.*

V. ang., var. *uncinata* Rouy
  NA, BBulls.

V. sativa L.
  mig.
V. tetrasperma (L.) Moench
  NFD (LaPyl.) *mig.*

V. hirsuta (L.) S.F. Gray
  NHarb (Wag). *mig.*

V. Cracca L.
  Com. n. to White Bay and IngornB. GrandL,
  PS, BC, GF, GI, StJ, *Bond, manive BT*
  V. BB

Lathyrus maritimus (L.) Bigel. *Everlasting Pea*
  Com. on coast. Barred I. WC, Baccalieu,
  FC, DR, BoTI, PaP. *Bug S+P. Bras, BB*

L. marit., var. aleuticus Greene
  Rae Bras Ral (Aug)

L. marit.,var. glaber (Seringe)Eames
   Torb,Arg,CH,BofI,StG, *99*

L. marit. var.
    BB.
L. palustris L.

L. pal., var. macranthus(T.G.White)Fern.
   Chimn,GGard,FC,MistC,StB, *Bard. Br. S.H. mid.*

L. pal. var. retusus Fern. & Srg.
    miq.
L. pal.,var. pilosus(Cham.)Ledeb
   Freq.on coast,inland toGF,CH,PS,WC,
Brig,FC, *Bard, Scr SxP, SavH.*

L. pratensis L.
   Herb. LaPylaie. *HumbR.*

LINACEAE

Linum usitatissimum L.
   NewHarb(Wag), *Shemonier (ama premiry)*

L. catharticum L.
   Humberm, *Han. Hund*

GERANIACEAE

Geranium pratense L.
   HarbGrace(McGill),BarredI,Fogo,Tw.,
BlackCove (ama) Bof Gr

G. Robertianum L.
   Tops, Snooks, SEA, Chimn, DeGratB, StP.
Brigus (Hilda Dove). BB.

G. Bicknellii Britton
   PlacJ, Gambo(Wag), GF, HH,

Oxalidaceae

Oxalis montana Raf. identif. a drawing of it call at Corner br,
Lang. Mrs Ayes sent for identif Gordon Winter of org.
   EUPHORBIACEAE in June, 1931 by Mrs. Ayres of Briefof. Dof & Card; fl.

Euphorbia Cyperissias L.
   HarbGrace, BC Brigus (Hilda Dove)

E. Heliossopia L.

E. peplus L.
   StP. CALLITRICHACEAE

Callitriche palustris L.
   Fogo, ApseyBeach, BeAC, HR, Steph, LitR,
Trep, Fogo Card StP. BB.

C. heterophylla Pursh
   Torb, CVL, MicCove, Tops, Kil, BJ, Whit, Plac,
CI, GF, RP, Badg, Quirp, Mauve, OW. Burp.

C. anceps Fern.
   Blomd, Bard, Burp

C. hermaphroditica L.
   Green'sHarb(Wag), DeerL, StB

EMPETRACEAE

Empetrum nigrum.L. *Blackberry*
    Com, StJ,Balena,OP,Fogo,BarredI,Baccalieu,
Chah,Blom,FC,TW,GGard,PaB,BBulls. *DogPen.Lmt²*
StP.

Ep. firma purpureum (Raf.)Fern.

E. Eamesii Fern. & Wieg. *Rockberry*
    Com. n. to NDB,e.Br,Humb and BofI.
StJ,OP,Baccalieu,Seem,Musgr,Blom,BluffHd,
PaB,Balena,Sagoni and BrunEt'(Wag). *Doct*
Quirp.LMt Buy StP. BB

E. atropurpureum Fern.& Wieg.   (Magd.)

? Corema Conradii Torr.   (Magd.)
    Nfd. acc. to Tuckerm.

AQUIFOLIACEAE

Ilex verticillata (L.) Gray
    NHarb(Wag),Glen,GF,RP,HareB(LaPyl),BP,
SL,GP,GrandL,Steph,

Nemopanthus mucronata (L.) Trel. *Brick. trudes Greanwood Catberry*
    Com. n. to NDB and E. Br,Humb. and BofI,
BC,Badg,Blom,Torb,Balena,C,MJ,Baccalieu,
GI,BofI. *Doct LH StP,BB*

*I, vert., var. tenuifolia (?39.) Wats.
    Lang. Indian Bridge (Ama)*

*I. vert. v. fastigiata (Bicku.) Fern.
    Indian Bridge (Ama).*

ACERACEAE

Acer spicatum Lam. *Sycamore White Wood*
    Com. n. to NDB.and BB. BB,WB,GF,Snooks,
Don,BoíI,StG,Mán,whít.Doct Stp. mig. BB

2
A. Saccharum Marsh
    BStG,acc. to Bell.

A. rubrum L. *Maple*
    Com. on MainI.no. toNDB and BB.
OP,BB,RP,BC,Badg, gg Buchan's Mine (Aud). Random
(aud)

BALSAMINACEAE

Impatiens biflora(L.) Walt. Fireweed
    Com. whit,Cl,NA,Brig,OldPerolle,BC,HR,
PaP,StG, Bara Stp, BB

RHAMNACEAE

Rhamnus alnifolia L'Her.
    Com. on n.coast and e. to Exploits and
Gander. GI,SP,GF,RP,Musgr,Brig,PS,BoíI.MC
Pac BB

MALVACEAE

Malva moschata L. *Brides & Grooms (some pink, some white)*
    StJ,BC,

M. rotundifolia L.
  BBulls,

HYPERICACEAE

Hypericum perforatum L.
  StJ,

H. ellipticum Hook.
  StJ,HarbBreton(Wag.)

H. boreale (Britton) Bicknell
  Com. n. to Exploits and BofI. StJ,BJ,
  Whit,BF,GF,BC,

H. canadense L.
  Com. n. to Exploits and BofI, Toro,Trep,
  StJ,Carb,BF,GF,StJ,SandyPt,PaB, *Burg,*
  StP,

H. virginicum L. *"Ground Honeysuckle" atSummit anh.*
  Com. n. to NDB and BofI, StJ,Kil,BJ,
  Carb,whit,GI,Lpt,TW,BF,BP,BC, *LW, Burg, StP,*

ELATINACEAE

Elatine minima (Nutt.) Fisch. & Meyer
  whit,BC, *Burg.*

CISTACEAE

Hudsonia ericoides L.
    Barred I,Baccalieu *StP.*

? H. tomentosa Nutt.
    "Les points culminans",LaPyl.in herb.

VIOLACEAE

V. cucullata Ait.
    Com. n. to NDB and BofI. Crabbs,Tops,
BC,Sum,GP,Fogo,Badg,FrenchmC,Whit,BJ,
Snooks,StJ,Tab,Ren,Balena GF,CI, *StP. BB*

V. cuc.,var. microtitis Brainers.
    Com. n. to Exploits. CI,GI, BlackI,TC,
GF,MJ,MaryAL,Chan,Balena,

V. nephropylla Greene
    GI,GF,Brig,IB,CH,Marb,Blom,HR,Tab,GGard,
mc, SC. SauC, Burn. Boat. Ed. Bard. 599. mik Pa'c Cru. BB

V. septentrionalis Greene
    BF,GF, *Crabbs(RBK)*

V. Selkirkii Pursh
    BofI,BC,HR,GeoP,PaP,GGard, *Doct. Quirp BB*

V. palustris L.   (Gaspé)
   Ed. Doct

V. pallens (Banks) Brainerd
   Com. GF, MJ, Gl, StJ, DR, Blom, FrenchmC,
Chan, Trép, Tc, Ed, Doct, LH, Buly, StP, Doct

V. incognita Brainerd
   Com. BC, Beč, PS, Gl, Balena, TC, CH, Steph,
Carb, BPS, Fc, arg, L mt, Doct TV

V. incog. var. Forbesii Brainerd
   Gl, Lpt, BF, GF, SB, WC,

V. renifolia Gray
   Burn.

V. renifolia Gray, v. Brainerdii(Greene)
                              Fern.
   FC, Marb, BofI, BC, Blom, Sc, Saue, Doct, Bč
Nat Pat, Doct

V. labradorica Schrank.
   Com. TC, Kil, BC, Blom, Look, BP, WB, Fogo,
BarredI, MJQ, Gl, BI, Fc, Coalk S9S, Oŭ StP,
Pač

V. arvensis Murr.
   StJ, StP,

V. tricolor L.
  DeGratB, *Miq.*

ELAEAGNACEAE

Shepherdia canadensis (L.) Nutt.
  On calc. rocks com. s. to NDB, Exploits
and BStG. GF, SC, FC, IB, CH, Wood I, Tab, StG,
Sgs, Bof. 99, B

LYTHRACEAE
Lythrum Salicaria L.
  NA,

ONAGRACEAE

Epilobium angustifolium L,
  Com. CH, BB, CP, GF, Don, Fogo, BC, StJ, GrandL,
Chan, BBules, MC, On, Miq,

E. ang., f. albiflorum (Dumort.)Haussk.
  watR, GF, GeoP,

E. ang., var. intermedium (Wormsk)Fern.
  GrandL, Quirp

E. ang., var. platyphyllum(Daniels)Fern.
  Doct   (Gaspé).

E. latifolium L.
  GF, BellI, Canada(Banks)Cape Rouge(LaP.l.)

128

E. latifolium L.
SC,FC,GrandL(Bell),GeoP,HR,HopeCove,StG,
Doyleïs, GF, Bell9(Canada B), Cap Rouge (La P.V). SavC.
BY. Burn. Cook. Boat. 593. Ha. Quirp. On.

? E. molle Torr.
   Pipestone Road acc. to Wagh.

E. nesophilum Fernald
   BBulls, StJ,Whit,Don,NHarb(Wag)BF. Doct. Eddy

E. Pylaieanum Fern.
   Trep, Harb Bret, Ramea, PaB. Burg

E. palustre L. Good-bye Summer (Pap)
Com. BarredI,FC,PaP,CH,GF,BJ,Whit,CH,PS,
Chan, SC, Brig, FC Doct, Quirp. Baid. On. StP,

E. pal.,var.monticola Haussk.
   Com. n. to ExploitsR and IB. StJ,Carb,
Whit,MaryAL, BBulls, FC, LW mig.

E. pal.,var. labradorieum Hausskn.
   CRay, LitR, Doct

E. pal.,var. manndjuricum Haussk.
   IB, Baid.

E. pal., var. lapponicum Wahlenb.
   Doct. Sav.

E. pal.,var. longirameum Fern.&Wieg.(Bab.)

E. wyomingense A.Nels.
    FC, Brig, CH, Ingorn, CStG, *Ed. Diet.*

E. davuricum Fisch.(Huds.Bay)
Big. Sav. SC, SJ9, Quirp, bc
P. *leptocarpum Haussk., var. macounii Trel.*
FC, Doct.
E. glandulosum Lehm.
  Com. on coast s. to Croque and BStG.
FC, Brig. CH, GGard, *2B Dich Boat.*   *Stp, mig.*

E. gland.,var. cardiophyllum Fern.
    StJ, Wat, PS, Musgr, *Bald. Bur.*

E. gland., adenocaulon (Haussk.)Fern.
    Com. Gl, DR, GF, Whit, MaryAL, WB, PaP, HR, Chan,
SN. FS, Burg. BB

E. gland., var. occidentale (Trel.)Fern.
    StJ, BJ, BC, *Trel.*

E. gland., var. perplexans (Trel.)Fern.
    DR, StG,

E. *brevistylum Barbey*
    mauve, Sav.

E. boreale Haussk. (Gaspé)
    Doct. Sav, Bald. ~~May~~ ~~Vb~~
E. *Behringianum* ~~Haussk.~~
E. ~~alsinifolium~~
    Mist. C

130 *E. scalare* Fern.
  Doct.

E. Drummondii Hausskn.   (Gaspé)
  Big Brk. Ed. Misk C

E. lactiflorum Hausskn.   (Gaspé)
  Doct. Quirp. Oh

E. alpinum L.   (Gaspé)
  Burn.

E. Hornemanni Reichenb.
  Com. s. to Croque and Cod Roy. FC, StB.
  CH, MuSgr, BC, Blom, Tab, Doyle's, Burn. Quirp. Doct.
  Cook BB

Oenothera *parviflora* ~~muricata~~ L.
  Tops, BF, GF, Badg, SEA, ShoalPt, CoalR,
  StG, QGard, ~~F9~~. St Lan, BB

O. biennis L.

O. perennis L.
  SoBildo(Wag), Gl, BF, GF, Humb, Steph, Lan,
  Curl, BB Gaudes L (to Holloway)

Circaea alpina L.
  Com. Chah, BB, FC, FrenchmC, BI, Lpt, PS,
  BC, Man, Salm, Tops, MC. Stp.

                    HALORAGIDACEAE

Myriophyllum alterniflorum DC.
  Whit, GF, FrenchmC, BC, HR, PaP, Sask, Yark, Oth P
  StB Miq Salmonier R (dwg) Chappells Arm (Avia)

M. exalbescens Fernald
PR, *Braun, FC Ottr P. Pac*

M. magdalenense Fern.   (Magd.)

M. verticillatum L.,var.intermedium (Magd.)

M. vert.,var.pectinatum Wallr.
GF,GeoP,

M. Farwellii Morong.   (Gaspé)

M. tenellum Bigel.
Treb,QVL,BJ,Whit,NHarb(wag),Gl,Lpt,FC,
RP,BC,

Hippuris vulgaris L.
Freq. Salm,BBulls,GF,Badg,FC,BC,Steph *Miq*
On

H. vulg.,var.maritima *(Hellan) Wahlenb,* Hartm. (Lab.Gaspé)

ARALIACEAE

Aralia hispida Vent.
Com. n. to NDB. and BofI. Don,OP,GF,Fogo,
BarrodI,Badg,Tops,BC, *Han*

A. nudicaulis L.
    Com. St.J, Tops, DR, GF, BB, BC, RopeCove, StB,
Beat, SC Lang.

            UMBELLIFERAE
Hydrocotllau . Barachois Br (RBK)
Sanicula marilandica L. var. borealis Fern.
    Gl, BF, GF, Badg, PR, BC, WC, HR, Tab, Beat, Doct, F.J.
Lang, PaC, BB

Osmorhiza obtusa (C.&R.) Fern.
    DR, FC, StB, BC, Bene, FrenchmC, Cooks'sBr, MC.
SavC, Ed, Doct, Baid, Humbm, PaC.

O. divaricara Nutt.   (Gaspé)
    Doct.

Cicuta maculata L.
    NHarb(wag)

C. bulbifera L.
    FC, HB, Steph, Beat. SavC

Carum Carvi L.
    Com.   St.J, Gl, BC, StB, Beat.

Sium suave Walt.
    BJ, Whit, NHarb(Wag), GF.

Pastanaca sativa L.
    StB, Ship Cove (Ama) BB.

Pimpinella Saxifraga L.
StJ(Will.)

Ligusticum scothicum L. ✓ *Alexanders*
   Com. on coast. Torb,Plac,Funk,WH,FG,PS,
BC,IB,Burgeo, *On.No⁰ StP.*

aethusa Cynapium L.
   StP⁰

Coelopleurum lucidum (L.) Fern.
   Com. on coast. Funk,Brig,IB,WC, *On. Ubo*
mig, Sort,

                    Health Root. Hell Trot.
Heracleum lanatum Michx. *Heltrope Embloch*
   Com. PlacJ,Salm,GF,BB,FC, *Burg. StP,*

H. Sphondylium L. *wild Parsnip*
   Trep!

Conioselinum chinense (L.) BSP.
   Com. BB,SEA,IB,WC,Badg,BC,StG,Tab,Snooks,
GF,Fogo,BarredI,Torb,FC,Brig,StB,IB,Burgeo,
FJ⁰No StP, Nan

C. pumilum Rose ✓
   Blom s.BBs, TC,Look,Stanth(Huntsm.) *No⁰*

C. *Gmelini (Bray) Stead,*
   Soct Bars, On

C. ~~Gmelini Torr&Gray~~. (Gaspé)
   Benthami (Wats.) Scu.

Angelica atropurpurea L.
   Badg,Conche(LaPyl.)FC,BC,BenC,SealI(B&G),
HR,Tab,Codroy, *BS*

A. *Laurentiana Fern.*
   Sort. On Gro, Sar H, PaC,

CORNACEAE

Cornus canadensis L. *Crackers, Crackerberry*
    Com. whit,Balena,DR,GF,WB,BelII,GrandL,StJ,
Fogo,BC,FrenchmC,Gaultois,PaB, *Burg.* *StP.*

C. can. var. intermedia Farr.
    FC,CH,PaB, *Bif. Doct. Hampdan(A.M.J).*

C. suecica L.
    Freq. on coast. DR,Wh,Quirpon,PaB,Chan,
Burgeo,CStG, *Sa᷃oC. Ha᷃. Doct. No᷃* *StP.* *Ral(Aug)*

C. stolonifera Michx. *Red Rod. Widdy*
    Com. GF,NS,Frenchmc,PaP,BB,GI,LitRodIndL,
WB,BC,StG,Tops,GrandL,StJ,PaP,whit, *Bard*
Fg᷃ Lang,
var. laurentiana Victorin
    Fc, Bof., PaP.,RedIndL, Glen PiC.
C. alternifolia . f.
    NA,BF,HH,BC, Benc,HR,RopeCove,Tab,Steph,
Lang,

ERICACEAE

Moneses uniflora (L.) Gray *Sweet Flower*
    Com. Tops,whit,Humb,Tab,BC,CH,BC,Sum,DK,TC,
arg,FC, *Quirp. Lang,*

Pyrola minor L.
    Com. n. to HareBay,E.Br.Humber and Bofl.
GF,FrenchmC,Sum,Doyle's,GeoP,BC,TC,BI,StG,
Sav,Burn,Quirp.On. TC Humb

P. secunda L.
 Com. n. to NDB,EBr.Humb.and BofI.
FrenchmC,DR,Don,Lpt,GF,MaryAL,OP,BenC,Musgr,
Fogo,BarredI,Tops,StJ,Salm,BBulls,PaB, *Doct*
Lang,
 P.sec.f.eucycla Fern.
 Doct,

P. sec.,var. optusata Turcz.
 Com. s. to NDB and BStG. DR,Lpt,TC,GF,FC,
PS,BB,Chim,BC,BenC, *Bard.*

P. chlorantha Swartz.
 StJ,GanderR,Baccalieu, Fogo,DR,Exploits,
TW,BF,LarkH,SerpR,StG, *Burin, Bard.*

P. elliptica Nutt.
 GI,NA,Lpt,BF,GF,

P. rotundifolia L. var. arenaria M&K.
 Man,Caro,WB,TC,GF,MJ,BP,LittleBlom(E&G),
FlatB(Bell), *Jg Lang,* (B13)

P. asarifolia Michx.
 Gl,BF,GF,Doyle's,

P. as.,var.incarnata (Fisch.)Fern.
 Gl,FC,Brig,PR, *MC. Sav. Bif. Burin, Bröad. Pac*

P. grandiflora Radius
 ? Tab,

Monotropa uniflora L. *Ghost plant*
    Com. DC,Don,GF,StJ,PaP,PS,Salm,Badg,
BofI,Brig,BB. *Doct. LMt Lang.*

M. Hypopithys L., *var (glabrous)*
    Freq. n. to NDB and IB. PS,SB,Badg,Snooks,
RP. *Doct, LMt BB*

Ledum groenlandicum Oeder, *Crystal Tea.*
                                    Indian Tea
    Com. BC,MJ,GF,LewisH,Blom,Balena,Fogo,
GrandL,Baccalieu,Chan,PaB,FrenchmC,Trep,
Doct.LN. Brand. Cook. StP, BB

L. palustre L.,var. decumbens Ait.
    Spooky records

                        False Honeysuckle. Bull's Eye
Rhododendron canadense (L.) BSP.
    Com. n. to NDB,EBr,Humb.and BofI. Q,MJ,RP,
GF,TC,Carb,Balena,OP,GP,BC,FrenchmC,StJ,
Baccalieu,BarredI,Chan, *Bart. LH StP, BB.*

R. lapponicum (L.) Wahlenb.
    BC,FC,Blom,Tab,BBs, *SavC, Burn. to Mid BB*

Loiseleuria procumbens (L.) Desv. *Mayflower*
    Baccalieu,Fogo,Croque,Seem,Balena,Push-
through and com. near w. coast.Blom,FC,PK,
Tab,Musgr,PaB, *SavC. Bif. Burn, Doct,Quirp,LMt,
Burg. StP, Ric BB*

Kalmia angustifolia L.
  Com. Steph,OP,GrandI,StJ,Gl,GF,Sum,IB,
Fogo,Baccalieu,Ch,Brig,PaB,FrenchmC,Trep,*Doct,*
Burn, Buzz StP.

K. polifolia Wang. *Gouldwithy*
  Com. Fogo,Baccalieu,BarredI,Whit,FrenchmC,
StJ,OP,Balena,Gl,MJ,PaB,BF,Q,Blom,Tab,
Musgr,Brig,Trep,*FC,Quirp, Buzz StP. Doct.*

Phyllodoce caerulea (L.) Bab.
    Blom,Mt.s.ofGr.CodRoy (Bell). *Dret RB*

Cassiope hypnoides (L.) D.Don
  Blom, *Doct*

C. tetragona (L.) Don.   (Lab.)

Andromeda glaucophylla Link.
  Com. Blom,Musgr,MJ,Fogo,Whit,FC,Badg,
GrandI,GP,Balena,OP,Blom,BF,PaB,Salm,GP,
RP,BB,Trep,PaB,Brig,*FC, Big Burn, LH, LMs, Ms*
StP, PdC

A. gl.,v. iodandra Fernald
    Tab,GGard,

A. Polifolia L.   (Lab.)

Chamaedaphne calyculata (L.) Moench. *False Whort*
  Com. Balena,MJ,Badg,RP,FrenchmC,Steph,
StJ,Fogo,PaB, *Rd StP, GN*

Epigaea repens L.
    Freq. n. to BB and e. to HermitageB.
BC, Blom, Tab, PaB, Chan, *L H. to, Lang. Lomond (Peggy Simpson)*
GP

Gaultheria procumbens L. *Mountaineer Tea*
    DeGratBay. HarbBreton (Wag). *S+P.*

? Arbutus Unedo L.
    Bonaventure (Acc. to Cormack)

Indian Hurt, Hardberry
Arctostaphylos Uva-ursi (L.) Spreng.
    Com. westw. GP, Blom, Musgr, Tab, Chimn, FC,
Brig, *Doet,*   *S+P. BB Ral (Ang)*

A. U., var. adenotricha Fern.&Macbr. (Mingan)

A. I., var. coactilis Fern.&Macbr.
    IB, GF, Fogo, BarredI, *to*

A. alpina (L.) *Spreng.* *Poison Berry*
    Com. on exposed crests, etc. Blom, FC,
Baccalieu, TW, BarredI, Seem, Musgr, Tab, Belena,
PaB, Brig, *Quirp L Mt. Burp, S+P, Mid. WB*

A. rubra (Rehder&Wilson) Fern. (Antic.)
Bif Burn, Cook, Brand, Norm. FC

Calluna vulgaris (L.) Hull.
  "head of St. Mary's Bay"(Cormack);
"Trepassey Bay, also very abundant"
(Cormack); "S.E.of Newfoundland con-
siderable tracts of it"(Cormack);
Renews (Harvey), Ferryland (Judge Robinson),
Caplin Bay (Harvey). "Terra Nova" (LaPylaie)
  (M. Costello)

Chiogenes hispidula (L.) T.&G. *Capillaire, Maidenhair.*
  Com. WB,GF,C,FrenchmC,StJ,MistC, *Tea Berry*
StP. Dobr.

Gaylussacia baccata (Wang.)K.Koch *Black Hurts*
  NHarb(Wag),BofI(E&G),LittleHarb(near BofI),
Steph,StG,PaB,LongHarb(Fortune), *StP.*

G. dumosa (Andr.)T.&G. var.Bigeloviana Fern.
  Trep,Whit,Steph,StG,PaB,Chan,LittleBay
(Fortune). *Burg Lang, Salmonier(Chuck)*

Vaccinium pensylvanicum Lam.   *Hurts*
  Com. Whit,DR,StJ,GF,BC,Musgr,GrandL,Fogo,
Baccalieu,Barred I,Salm,GGard,BBulls. *LH. FJ*
PdC

V. pen.,var.myrtilloides (Michx.)Fern.
  NDJunct,BF,C,StG,

                              Tobacco Hurt. Sweet H.
V. pen. var. angustifolium (Ait.)Gray
  Com. BC,TC,Baccalieu,C,Balena,TW,MJ,StJ,
Plac,StG,GP,Chan, *TC Save Bront. FJ. Lnot.*
StP, Dobr. Ral(Wing.)

V. canadense Kalm
  Cape Broyle

140

V. uliginosum L., var.alpinum Bigel. *Ground Hurts Bilberry*
   Com. Prep.OP,FC,PaP,StJ,Chan,Fogo,TW,GF,
   BP,Baccalieu,C,Balena,Look,Musgr,PaB,Brig,
   IB, *Doct, Quirp, StP, Pac Bg*

V. cespitosum Michx.
   St.Anth,SavC,FC,BC,Blom, *Burn. Boat, Doct Quirp*

V. nubigenum Fern.
   Doct.

V. ovalifolium Sm. *Blueberry Hurt. Mathers*
   Snooks,WhiteB(Wag),Croque(Banks,PaPyl.),
   BC,BenC,FrenchmC,Blom, *Boat, Doct,*

V. Vitis-Idaea L., var.minus Lodd. *Partridge Berry*
   Com. Musgr,IB,KB,BB,DR,StJ,C,MJ,GF,FC,
   SL,Fogo,Baccalieu,BBulls,Brig, *Lnd StP,*
   Nak

V. Oxycoccus L. *Marshberry*
   Com. BJ,Doyle's,Blom,StG,MJ,TW,FrenchmC,
   Gl,DBn,GF,Balena,BBulls,IB, *LH StP,*

V. Oxyc va,
   Pac

V. macrocarpon Ait.
   Com. n. to Croque, E.Br.Humb.and BofI.
   BBulls,Torb,StJ,Krl,Carb,OP,NHarb,Gl,BP,
   PP,BP,StG, *StP,*

DIAPENSIACEAE

Diapensia lapponica L. *Moss Lily. Ground ivory flower*
Kantem(Wag), Croque(LaPyl), HH, Look,
Seem, Musgr, Blom, PaB, Chan, Balena, Herb Bret(Wag)
Doct. Quirp. & Mt. Burg. StP. BC

PLUMBAGINACEAE

Statice labradorica (Wallr.)Hubb.&Blake
SC,

S. lab., var. submutica Blake
SC, PR, BBs, Blom s, Coaln, LewisHills. *SavC. Big*
Cook, Braid, Boat 4mi, 5J5. to Pab

Limonium Nashii Small, var. trichogonum Blake
BStG(LaPyl. as n.sp. in mss.)   Nfd(Miss
Brenton in herb Hook.)

P. veris L.
StP.   PRIMULACEAE

Primula ~~farinosa~~ *Laurentiana Fern* L. (Lab.)
SC, PC, *Qf mid PaC BB*

P. ~~farin., var. americana Torr.~~ (Ocspe)
StB,

P. ~~farin., var. incana (M.E.Jones)~~ Fern.
PC, *Burn, Braid, York, Dap*

P. farin., ~~var.macropoda~~ Fern, *Salmon Flower*
Com. s. to NDB and BStG. TW, TC, FrenchmC,
CH, PS, PaP, PC *Norin, Doct. Baird, F.J.*

P. borealis Duby (Arct.)

P. egaliksensis Wormsk.
FC, StB, *Bir, Cook, Brand, Boat, Ed, Sg3, On, Pac,*

P. ~~sibirica Jacq. var~~ arctica Pax.
Big, Cook, Norm, Watts B.

P. mistassinica Michx.
  Com. s. to GanderR.and ExploitsR, E.Br.
Humb.and CodRoy. Tab, Doyle's, FrenchmC, GF,
MJ, GrandL, MiddleA, Gl, BC, Blom, IB, FC, Brig,
SavC, L Mt, mid Pac, BB

Androsace septentrionalis L. (Gaspé)
  BB

A. Chamaejasme Host (Arct.)

Samolus floribundus HBK. (PEI)

Lysimachia punctata L.
  StJ, BC.

L. terrestris (L.) BSP.
  Com. n. to Exploits R. and BofI. Steph,
Carb, Kil, Tops, Whit, *Gov, Bur, StP, Mig, BO*

L. nummularia L.
  Harb Grace (McGill)

L. thyrsiflora L. (Magd.)

Trientalis borealis Raf.
Com. StJ,Fogo,Baccalieu,Balena Brig,
LewisH, FC. Quip. LMt StP.BB

Glaux maritima L.   (NB.)

G. marit.var.obtusifolia Fern.
DR,NA,HareB(LaPyl),Brig,PS,CH,StG(LaPyl)
PaP, Bard. Ral. Sav. Wo. Burp.

Anagallis arvensis L.
HarbGrace(McGill) mig.
a. tenella L. (StP.)

Centunculus minimus L.   (SableI,PEI)

OLEACEAE

Fraxinus nigra Marsh.
SL,PPS,GP,DeerL,HumbR,BofI,BenC,HR,
CoalR,BenoitBr. mid BB

GENTIANACEAE

Gentiana nesophila Holm.
Com. s. to Canada Bay and BStG. MistC,
FC,StB,Brig,OldFerolle,IB,CH,PaP,GGard Bard.
Sg9, OttsP.Seil Sd, Edbpi BB Cladhis (RBC)

G. Amarella L.
Com. on w. coast s. to BStG. Quirpon,
MistC,FC,StB,Brig,PS,Chimn,StG. On, Bard.Sav.

G. propinqua Richardson
   StAnth,Quirpon,MistO,FC,StB, InfornB. *Big, Burn
Rae, Doct, OttrP*

G. nivalis L.   (Lab.)

Lomatogonium rotatum (L.) Fries
   Com. s. to Canada B. and Brig B. Quirpon,
Conche,FC,StB,Brig,MistO *On, Bard,OttrP, SurC*

Halenia deflexa *(Sm.)* Griseb.
   Com. near coast Salm,Carb,Torb,Tops,OP,FC,
GH,StG,CapeStG,PS,StJ,BarredI,Chimn,Bisc,
IB,BBulls,Gaultois. *Quirp, On, Bay, J-9", StP.*

Bartonia paniculata (Michx.)Rob.,
   var. iodandra (Robins.)Fern.
   Tre B,Bisc,Holyr, Q,GrandL,BC,StG,Steph,
PaB, *Lfl, LMt, Bay, BurntIsl, OB*

B. virginica (L.) BSP.
   StPh
Menyanthes trifoliata L. *var. minor Michx.*
   Bog Bean
   Com. FranchinC, BB,PS,GF,MJ,Look,Badg,
FC,PaB, *StP. PaC*

Nymphoides lacunosum (Vent.)Kuntze
   hP

APOCYNACEAE

Apocynum androsaemifolium L.
  BF,GF,Badg,BP, *Gander Lake (Robin Reid*
                                no. map)

A. medium Greene
  BF,GF,

A. cannabinum L.
  GF, Badg(Bull),

CONVOLVULACEAE

Convolvulus sepium L., v. americanus Sims.
  Tops(Wag),Spreadeagle(Wag),NA,Botw,SEA,
  Steph, *Wo*

  C. sep, var. pubescens (Gray) Fern.
    S+P,

POLEMONIACEAE

Polemonium boreale Adams   (Arct.)

Collomia linearis Nutt.   (Gaspé)

BORAGINACEAE

Lappula echinata Gilib.
  St.J,GF,BC,

Hackleia deflexa,v. americana(Gray)Fern.&
                                Johnst(Gaspé)

Omphalodes linifolia (L.) Moench
   Nfd (Morison acc. to Gray)

Symphytum officinale L.
   StJ,HarbGrace(McGill).

S. asp.
   Colinet & Witchhazel Road (Ama)
Lycopsis arvensis L. (Gaspé)

Myosotis scorpioides L.
   StJ,HarbGrace(Trapnell),BC,Humb, Bard. M

M. laxa Lehm. ✓
   Man,StJ,Kil,BJ,Carb,Whit, BB

M. arvensis (L.) Hill
   StJ,Carb,Plac,Trep, StP.

Mertensia maritima (L.) S.F.Gray Ice Plant
   Com. on coast, FC,IB,CH,DR,Frenchmc,
FunkI,Tops,BarredI. Wo, Burg, StP.

   Borago officinalis L.
      StP.
               LABIATAE

Scutellaria lateriflora L.
   RP,DeerL,Humb(E&G).

S. epilobiifolia A. Ham. Red Tops
   Freq.IB,Humb,Don,Whit,StG,RP,BF,DR,BI,
Cl,Whit,BarredI,Irishtown,Man,FC,Brig, Bard
StP, M

Nepeta Cataria L.
  John ᴮᴮᵒᵘᶜʰ,

N. hederacea (L.) Trev.
  StP, *Humbh,*

Scarlet Runner

H. hed., var. parviflora (Benth.)Druce
StJ, PaP,

Dracocephalum parviflorum Nutt. .(Gaspé)

Prunella vulgaris L.
  GF, BC,

P. vulg., var. lanceolata (Bart.)Fern.
Marb, PaP, StG, *StP, BC*

var villma

Galeopsis Tetrahit L.
  StJ, Forb, Tops, BC, StB,

G. Tet., var. bifida (Boenn.)Lej.&Court.
Com? StJ, Caro, GF, OP, Tops, BofI. *Bard, StP,*
BB

G. Ladanum L.
  StJ(Wag) *StP,*

Lamium amplexicaule L.
  StJ, NHarb(Wag), *StP,*

L. purpureum L.
Mauve, Burg. StP. BB(Benson)

L. <u>hybridum</u> Vill.
  Com. neare. coast. Prep,BBulls,StJ,Carb,
BarredI, *Buie. StP.*

<u>Stachys palustris</u> L.
    StJ,Tops,Kil,HarbGrace,Carb,SandyPt(Wag),

<u>Satureja vulgatis</u> (L.) Fritsch.
  Salm,GF,Chimn,Marb,Beno,GeoP,GGard,Tab,
PaP,DeGratB, *Doct, BB*

<u>Lycopus uniflorus</u> Michx.
    Com. n. to NDB, M.Br.Humb.and BofI.,StJ,
Carb,BarredI,BBulls,BF,GF,BC, *Buie StP, MB*

L. <u>americanus</u> Muhl.
  Salm,Whit,GF,RP,CodRoy(Bell).

<u>Mentha cardiaca</u> Gerarde
  BC.

M. piperita L.
  StP.

M. arvensis L.
StJ,Man, Carb,BBulls,StG,RomaineBr. *Doct, StP,*
MB

M. arv.,var. glabra Benth.
  HR, *BM*

M. arv.,var. canadensis (L.) Brig.
Freq. n. to NDB and BrigB., Don, Tops
WB, Lpt, BF, Brig, Chimn, PaP, whit, *Rock Burg,*
StP, BB

M. arv.,var. glabrata (Benth,)Fern.
Report by Wag.

SOLANACEAE

Leucophysalis grandiflora (Hook.)Ryd.
(Gaspé)

Solanum Dulcamara L.
StJ, Tops,

SCROPHULARIACEAE

Verbascum Thapsus L.
BB,

Linaria repens Mill.
Ren, StJ, Torb, BBulls, whit, *Curl*

L. sepium Allm.
StJ, Tops(Wil.)

L. vulgaris L.
BBulls, Torb, StJ, whit, HarbGrace(Wag),
LaScie(Wag), SpruceBr, HarbBret(Wag), *p/s*

Scrophularia nodosa L.
  Meadows, WC, Humb, BC,
  Murrays Pond & Beachy Cove (Awa). Ham,

Chelone glabra L.
  Badg, Ferryl, Whit, LH mig.

C. gl., var. dilatata Fern. & Wieg.
  Cl, Whit, GP, NA, Steph, BC, StG,

Mimulus moschatus dougl.
  NA, BC, BenC, LitR, Windsor Lake (Awa) Hanbm BB,

Limosella aquatica L.
  HarbBret,

L. subulata Ives.
  Carb, NA,

Gratiola aurea Muhl.
  Whit,

Digitalis purpurea L.
  GeoP, Salmnier (Awa)

Veronica longifolia L.
  QVL(Wil), Cl,

V. americana Schwein.
  HawkesB(Wag), CH, LH Curl w. Humbm BB

v. scutellata L.
  StJ,Whit,Diläo(Wag),Salm,BF,GF,RR,DeerL,
  Long.

v. officinalis L.
  StJ,Tops,Kil,Doyle's. StP

v. off.,var. ~~Journeefortii(Vill.)Reichem.~~
  StJ,Tops,Salm,BBulls,Whit, Curl.

v. alpina L.,v. unalascensis C&S.(Gaspé)
  Doct.

v. serpyllifolia L.
  Com. n. to NDB and IB. Snooks, MaryAL.
  StJ,BI,Carb,BC,Lpt,Chimn,PS,Cook.Burg. StP.

v. humifusa Dickson,
  BC,WoodsI,BenC,Humb,Steph,BB, Doct,On

v. arvensis L.
  Carb,Plac(Wil),CI,TC,Trep, StP,

v. agrestis L.
  StJ,Carb,BBulls. StP,

v. persica Poir. Glory-in-The-Mornin
  StJ,Carb,Trep,BBulls,BC,HarbBret,

Castilleja pallida (L.)Spreng.,v.septentri-
  onalis (Lindl.) Gray
  Com. s. to Croque,E.Br.ofHumb.and BofI.
  FC,StB,GP,GrandL,Musgr,PS,CH,BB,Chimn,
  ShoalPt,Blom, Bard.Har BB PaC

Melampyrum lineare Lam.
    GP,StG,PaB,CStG, *Lang.*

Euphrasie Oakesii Wettst.
  FC,StB,SC,Brig,IB, *Quirp. On. FS. BB*

E. Williamsii Robins.
    Quirp. On BB

E. Williamsii Rob., v. vestita Fern.&Wieg.
  Look,Blom, *SH*

E. leucantha

On
E.
    Tab,   *E. fil.*

E. purpurea Reeks
  Com. on w. coast n. to IB. PS,CH,SEA,BofI,
Steph,SrG,PaP,CStG,PaB, *LN. Buy WB BB*

E. purp.,forma candida Fern.&Wieg.

E. purp.,var.Randii (Rob.)Fern.&Wieg.
    Arg,Naultois,Ramea,Burgeo,PaB, *FS Brig. LH*

E. purp.,var. Randii,f.albiflora (Rob.)
                Fern. & Wieg.
    Kil,Trep,WH,

E. purp.,var. Farlowii (Rob.)Fern.&Wieg.
    Torb,Barred I,Steph,StG,GGard,PaB,Trep,
StJ,Arg,

E. purp. var. Farl, f. iodantha Fern. & Wieg
    Burg.

E. disjuncta Fern. & Wieg.
  Com. on w. coast & e. to NDB. BI, BF,
GF, Q, KB, FC, IB, PS, CH, BB, Look, SEA, Blom, BC,
Tab, HK, Brig, StB, *Bard, Sgg. Ha Pac*

E.   (pink)
  FC,

E. arctica Lange
  Tab, StAnth, Quirp, *SH Pac*

E. canadensis Towns. (Gaspé)

E. stricta Host.
  StJ, Torb,

E. str., var. tatarica (Fisch.)Fern.&Wieg.
                                    (Gaspé)

E. americana Wettst.
  Com. n. to NDB and Ingorn.B. Carb, Whit,
Kil, Fogo, Trinity, DR, RF, GI, BF, Torb, GrandL,
Humb, FrenchmC, BC, Doyle's, Chan, PS, BB, CH,
Gaultois, OP, *Sgg, Bard, LH, Bury, StP, BB*

Bartsia alpina L.   (Lab.) *Velvet Bells*
  Cook, Brand, Norm. Boat, Half-way

Odontites rubra Gilib. (PEI)

154

Pedicularis lapponica L.   (Lab.)

P. sudetica Willd.   (Arct.)

P. arctica R.Br.   (Arct.)

P. hirsuta L.   (Arct.)

P. lanata C.&S.   (Arct.)

P. euphrasioides Stephan   (Lab.)

P. groenlandica Retz.   (Lab.)

P. flammea L.   (Gaspé)
Sav. Biz. Half-way.

P. capitata Adams   (Arct.)

P. palustris L.
    StJ, Waterford—PettyHarb.(Harperized)
Black Marsh Road (Ama)

P. sylvatica L.
    StJ, MidCove, Don, Carb, Waterford-Petty-
Harb (Harperized) *Murray; Pond (Ama)*

Rhinanthus Crista-galli L.

R.C., var. <u>fallax</u> (Wimmer&Grab.)Druce
  Com. weed. StJ,Badg,RP,BarredI,GrandL,
OP,Humb,FrenchmC,StB,BBulls, *S+P*.

R. borealis (Sterneck)Chabert.
  Com. s. to CH. FC,CH,IB,PS,Brig, *SH*

R. Kyrollae Chabert
  NDB and Gander R. to BofI. DR,GI,BF,GF,
BI,Maro,BC, *79*

R. se ⁱ   *R.*
                    BC /

R.

    Tab,GGard,

                L'ENTIBULARIACEAE.

Utricularia geminiscapa Benj.
  Freq. n. to BofI and Exploits R. PaB,BC,
StG,C, *Bard, Burg 1918*

U. vulgaris L.
  Freq. n. to NDB and BofI. Carb,Whit,BC,

U. vulg.,var. americana Gray
  Plac(Wil),GF, *Mig*.

U. minor L.
  Freq. Prep,Carb,Lpt,GF,C,FC,Blom,FrenchmC.
SgS,OtterP. Mig. PaB 1918

U. intermedia Hayne
 Com. 2 JunctBr, Gl, TW, BJ, Sum, Blom,
BarredI, CoalR, TopS, PlaC, BluffHd, FC, Trep *LW*.
Mig, PaC

U. cornuta *Michx*.
 Com. n. to NDB and BB. BB, RP, Carb, Look,
BC, StG, GrandL, WB, Badg, Salm, OP, Trep, PaB,
Bard, LW, Burg, GtP,

Pinguicula vulgaris L.
 Com. FC, MistakenC, BB, FrenchmC, Blom, BC,
CH, BP, GrandL, Q, GF, Gl, BarredI, Torb, PaB,
Burn, BodT, Bard. SfP, Nac, mid, BgP

P. villosa L.   (Lab.)

P. alpina L.   (Lab.)

OROBANCHACEAE

Orobanche ~~uniflora L.~~ *terrae-novae Fern,*
 BC, MarB, BenC, Tab, Dildo(Wag), GF, Gl, *Bard*
SfB, Burn, DocT, 79 Lang, Nac, mid, BgB

PLANTAGINACEAE

Plantago major L. *Rat-tail*
 Com. FrenchmC, BC, Carb, GF, *SfP, BgB*

P. major, var. asiatica (L.) Dcne.
BgB

P. maj.,var. intermedia (Gilib.)Bcne.

P. eriopoda Torr.   (Gaspé)

P. juncoides Lam.,v. decipiens(Barnéoud)
                              Fernald
  .Coast n. to NDB and BofI. StJ,Snooks
BC,FrenchmC,Chan,Gaultois,Burgeo,Torb,*Bard,*
On. 79, Burg StP.

P. junc.,var. glauca(Hernem.)Fern.
  Com. on coast. Fogo,Baccalieu,Chan,
FC,Tab,*Burn. Ha. S95.On* *Pac*

P. junc.,var. laurentiana Fern.
  Baccalieu,Funk,Plac,CH,Steph,*No. 79*.

P. oliganthos R.&S.   (Gaspé).

P. olig.,var. fallax Fern.
  Com. on coast. FC,FrenchmC,LitR,Kil,
NA.*Bard. On. SavC. Burgo BB*

P. lanceolata L.
  StJ,BC,*CAP, BB*

P. lanc., var. sphaerostachya M.&K.
  BC,StJ.

P.
curl
Littorella americana Fern.
  QVL(Wil),Whit,Plac(Wil),Badg,

Galium Aparine L.
  Spook records

G. kamtschaticum Steller.
  SB,BC,BenC,Blom,GreatCodRoy(Bell).FrenchmC,
  Doct 86

G. Mollugo L.
  StJ.

S. saxatile L.
  Trep.

G. palustre L.
  Com. n. to NDB and IB. GF,Gl,Don,CVL,
  CH,PS,BC,FrenchmC,Tops,StJ,BBulls, *Lang.*

G. ulig?
    BB

G. trifidum L.
  Freq. n. to NDBand BStG. BF,RP,DR,
  whit,Carb,StJ,GGard,Salm,whit,*Na,Doct,Bard*
  Burg

G. trif.,var. halophilum Fern.&Wieg.
  Com. on coast. Cl,BarredI,SEA,PS,
  Steph,StG,Trep,Brig, *Ral*

G. Brandegei Gray   (Lan.,Gaspé)
  mumve

G. Claytoni Michx.
Freq. n. to NDB and BofI. BF,BJ,Gl,Don,
Carb,Lpt,RP,Whit, *FC. Lang.*

G. Gl.,var. subbiflorum Wieg.
PaP,BC, *MM*

G. labradoricum Wiegand
Freq. s. to Gander R. Exploits and CapeRay.
Gl,Lpt,BF,GF,GP,CH,PS,BC,LarkH,PaP,StG,Brig,
FC, *Goat, Rat, Ed, LH, StP, Pat*

G. asprellum Michx.
Freq. n. to NDBand BofI. StG,PaP,GF,Whit,
SalM.Humb, *LH, BB*

G. triflorum Michx.
Com. BJ,TC,Lpt,GF,Badg,PS,BC,Blom,FrenchmC,
Brig,StB, *Bear, Mauve, Na. Lang. MB*

Mitchella repens L.
FlatBay Br(Bell),PaB *Lang.*

*Houstonia Faxonorum (Pease&Moore) Fern.
StP.*

CAPRIFOLIACEAE

Diervilla Lonicera Mill.
Com. n. to NDB.,Exploits and BStG. *GF, Stg,
Tops Mig*

Lonicera villosa (Michx.)R&S.
Freq. esp. on w. coast. SC,FC,Brig,IN,Torb,
Bart, 3JS, DogPen, Yank, StP, Pat

L. vill., var. Solonis (Eaton)Fern.
  Humb, GF, *Big*

L. vill., var. calvescens(Fern.&Wieg.)Fern.
  Freq. Fogo, BarredI, HopeC, GF StJ, Whit,
GP, FrenchmC, MistC *Burn, OR, Sav, Lang*

Linnaea borealis L., var.americana(Forbes)
  Ground Ivy, Trumpet Flower   Rehder
  Com. GF,DK, BC, FrenchmC, GI, Doyle's, Humb,
Char, Tops, StJ, MistC, *Burn, StP, Har*

                                    White Wood
Viburnum Opulus L., v. americanum (Mill.)
Trash Berry, Joint-wood Berry   Ait.
    Freq,n. to Exploits, Humber & BofI. KB,
GF, Badg, GI, RP, BC, *Hunt BB*

V. pauciflorum Raf. *Squashberry*
    Com. RP, FrenchmC, BB, GI, GI, HR, WB, BC, Man,
Tops, StAnth, GeoP, *FC, Quirp, Doct, FS, Burg, Doct*

V. cassinoides L.
  Com. n. to NDB, E. BrHumb.and BofI. StJ,
OP, WB, Tops, BC, BBulls, GrosMorne *PS, Lang, StP,
Op*

Sambucus racemosa L.
    Com. n. to NDB and BB. BB, FrenchmC,
PlacJ, BC,

 *S. canadensis (or is it nigra?!
    (Ault)*

*Symplocarpus racem. v. laevig.
   Self seeding & spreading, Beachy Cove (Ault)*

ADOXACEAE

Adoxa Moschatellina L.

VALERIANACEAE

Valeriana sylvatica Banks
 Croque(Banks), BC, HR, Tab, GGard,

V. officinalis
  Brigleshall (Ama)

DIPSACACEAE

Knautia arvensis (L.) T. Coult.
 StJ, BBulls, *Brigus (Hilda Dove);*
   very abund, infield

Campanulaceae

Campanula rapunculoides L.
 StJ, Carb, CI,

C. rotundifolia L.(Complex; needs watching)
 Com, Tops, Humb, Bogo, StJ, GF, FranchmC,
Chan, Humb, FC.Gaultois. *Bard, Sgg. Quirp, Fg,
mig, stp, Pac*

C. rot., var. alaskana Gray
 IB, *Bard Pac*

C. uniflora L.   (Gaspé)

LOBELIACEAE

Lobelia Kalmii L.
    GF,IB(LaPyl.).MiddleA,GooseA,Humb,BC,
Blom,HR,Tab,StG,Steph,SealRocks(Wag),BB.

L. Kalm.,var.
    OStG,

L. Dortmanna L.
    Com. n. NDB and BB. BB,FrenchmC,WB,
BF,Carb,Torb,RP, TC,BC,Salm,Fogo,BarredI,
PaB,Trep, *miq*,

COMPOSITAE

Eupatorium maculatum L.
    Salm, *LH Topsail & Princis Look Out (Ama) BB*

E. mac. var. foliosum (Fern.)Wieg.
    Man,BJ,GF,BluffHd,PaP,HR, *Bard, Doct, Gov,
BB,*

Solidago hispida Muhl.
    Com. near w. coast & e. to NDB, FC
FrenchmC,LarkH,Goose, SEA,IB,Tab,Blom,TC
Snooks,GF,Fogo,BarredI,CodR,BC, *Bard, Doct,
Cook, Norm L Mt, Eddy,*

S. hisp.,var. lanata (Hook.)Fern.
    Tab,

S. hisp., var. disjuncta Fern.
  SB,

S. hisp., var. tonsa Fern.
Look, Blom, BBs, *No*

S. hisp., var. armoglossa Fern.
  Apsey Beach(Wag), Meadows(Wag.) *Jg BB*

S. hisp. var,   *var*
     BB   *BB*

S. multiradiata Ait.
  Freq. s. to NDB and BStG, SC, FC, IB, PR,
BB, Blom, Tab, Fogo, Lewis H. *Burn, Doct, Sgg, On.*
Anch, Mauve, Norm, Mil, Pas, BB

S. macrophylla Pursh.
  Com. Lpt, Snooks, FC, Look, BC, Salm, OP,
WB, Goose, Tops, BofI, Chan, Chimn, StJ, FC, Gault-
ois, *Jg mig*

S. mac., var. thyrsoidea *(E. Meyer) Fern.*
  FC, Blom, Gros Morne, GGard, *On, Doct, Quirp.*

S. calcicola Fern.
  GF, HR, *BB*

S. menalis Fern.   (Gaspé)

S. chlorolepis Fern.   (Gaspé)

S. ~~nanilis Pursh~~ (= S. uliginosa Nutt.)
  Com. on w. coast and e. to Exploits.
PaP, GrosMorne, Look, Blom, BC, IB, TC, BF,
Badg, GF, Barred I, Humb, FC *Quirp, Mauve, Doct F9*
Gov Ha

S. ~~hum~~ *ulig.*, var. peracuta (Fern)
  BP,

S. sempervirens L.
    SEA, HarbBret(Wag), *Wo, SN.*

S. uniligulata (DC.)Porter
   Com. n. to Trinity Bay and BofI. StG, Cl,
Badg, Whit, Blom, LitR, PaB, Burgeo, Trep, 3Bulls,
LH Mig BC
  S. unilig., var. neglecta (PtG.) Fern.
               Crabbes (RBK)
S. rugosa Mill.
   Q LH. FJ

S. rugosa, var. sphagnophila Graves
    WC,

S. rug., var. villosa (Pursh) Fern.
    Com. n. to NDB and BB, BC, Cl, TC, HR, GF, *StP,*
BHo

S. lepida DC,
  TC, *Doct.*

S. lep.,var. fallax Fern.
  Freq. n. to Exploits R and HarrysR.
Tops,GF,RP,HR, *Bafg*

S. Bartramiana Fern.
  GF, *BB*

S. canadensis L.
  GF, Goose Arm, *Hav*

S. graminifolia (L.) Salisb., var.Nuttallii
                    (Greene)Fern.
  Whit,Cl,

S. gram.,var. septentrionalis Fern.
  BF,GF,

Aster radula Ait,
  Com. eastw. and n. to Exploits R. WB,
GF,Salm,Cl,Trep,   *StP, BB*

A. rad.,var. strictus (Pursh)Gray
  Com. BF, BB,GrosMorne, TC,Carb,G,NA,
Man,StJ,GrandL,Badg,StG,Torn,Fogo,BarredI,
Look,BC,FC,MistC, *Quirp, OK, LHa, 79, BB*

  A. saxetilis (Fern.) Blanchard
    Murray's Pond (Aua.)
  A. longifolius Lam.
  Tops,Meadows,LitHarb,WG,GF,Badg,

  A. paniculatus Lam.
    Crabbes (RBK).
  Bellis perennis L.
    Lang, Curl.

A. novi-belgii L.
  Com. Coal R,FrenchmC,SandyR(BStG),Torb,
StG,Blom,BB,SEA,RP,TC,Snooks,BBulls,
Gaultois. *Wo. Gov. No Burg Hd*

A. foliaceus Lindl.
   Com. StG,BC,GP,RP,Whit,RP,HR,Tab,SEA,BB,
RomaineBr,Blom,BluffHd,IB, *Doct. Dog. Bard.*

A. fol., var. frondeus Gray
     TC, *Dog. Yank*

A. fol., var. subgeminatus Fern.
   Tab, *Crabbes(R.&k).*

A. anticostensis Fern.   (Antic.)

A. puniceus L.
   Freq. n. to NDB and BB. BC,GrosMorne,
Salm,BBulls.

A. pun., var. firmus (Nees)T.&G.
   Com. n. to NDB and BofI. Snooks,BC,TC,
Torb,GP, *Otter. LH Hd*

A. pun., var. oligocephalus Fern. *Pitnages*
   WB,StAnth,Quirpon,ShoalPt,Blom,CoalR, *Doct.
Hd*

A. umbellatus Mill.
  Com. n. to NDB and BB. TC, Snooks, BB,
GF, BC, HolyT, Torb, Grand L, BluffHd, StG, Tops,
LitR, BBulls. FB Burg Mig.

A. linariifolius L.
  (Cormack acc. to Hook. -- "OK"-Blake).

A. acuminatus Michx.
  PaB, Xmt.

A. nemoralis Ait.
? Com. Carb, Cl, Q, TC, BC, Grand L, WB, Barred I,
Fogo, StJ, Salm, Chan, OP, Burg Mig. BB

A. nem., var. major Peck,
PaB,

A. laurentianus Fern,   (Magd.)

Erigeron hyssopifolius Michx.
  Com. s. to Exploits R., and CodRoy. FG,
BofI.SB, Tab, BIon, GF, Musgr, PS, CGard, SC, 4mi
4mi, OP

E. hys., var. villicaulis Fern.
  Tab, FC 4mi. Hd. Ice SC PaB

E. uniflorus L.   (Lab.)

E. compositus Pursh, var.
                    (Gaspé)

E. alpinus L.

E. ramosus (Walt.)BSP.,var.septentrionalis
                    Fern. & Wieg.
   BF,GF,RP,Badg(Wag),DeerL,HR,StG(Howley),

E. acris L., var. asteroides (andrz.)DC.
                    (Gaspé)

~~*borealis (Vierh.) Simmons*~~
E. ~~acr., var. oligocephalus Fern. & Wieg.~~
   N, Mauve Quirp.

E. lonchophyllus Hook.   (Antic.)
E. canadensis L.
      StP.
Antennaria eucosma Fern. & Wieg.
   Tab,GGard,*Big, Burn,Cobk, Brand, Norm, Boat 4m.*

A. cana (Fern.&Wieg.) Fern.
   Tab,PR,FC,Brig, *SC, Sav, Big, Boat, Burn, Brand,
Cook, Norm, Doct, St-J.J. Pav*

A. straminea Fern.
   Tw,Brig,PR,GGard, *SavC, Burn, Shoal C, SJJ, Ice
Hau Big be'*

*A. vexillifera Fern.
   Cook, Boat Pav.
A. Longii Fern.
   Brand Norm Boat, 4 mi Bob*

a — ?, Has BB   *a – ? Pac*

A. albicans Fern.
   Tab, *Burn, Cook, Brand. Ices.*   *A. nsp - BB (Jook)*

A. alpina (L.) Gaertn.   (Lab.)   *a, nsp. BB (Lind Cha)*
a. wiegandii Fern.
   Sgg. Pac
A. pygmaea Fern.   (Lab.)

A. spathulata Fern.
   Holyr, RP, PR, Tab, GGard, *Burn. Sgg. LN. LL mt*
Pac
   a. spath. var. continentis Fern. v St J.
A. petaloidea Fern., var. subcorymbosa Fern.
   Burn
   BF,

A. glabrifolia Fern.   (Lab.)

A. neodioica Greene
   StJ, RenR, Kil, GF,

A. neod., var. attenuata Fern.
   BF, GF,

A. neod., var. gaspensis Fern.
   Tops(Wil), SB, *LN Sg*

A. neod., var. chlorolepis Fern.
   StG,

a — ? umbicaps   *a. castanea*
   mid   *Pac*
a fuscescans   *a. ellipa.*
   mid.

A. neod.,var. rupicola Fern.
  Carb,High Gregoi,TC,GF,

Anaphalis margaritacea (L.) B.&H.
  Avalon to Exploits. BBulls,CI,GF, *LMt.*

A. marg.,var. subalpina Gray
  Trep,Bisc,StJ,Holyr,Carb,BB,Blom,StG, *On*
Bard Lang, BB

A. marg.,var. occidentalis Greene.
  Com. n. to NDB and BofI. CI,Fogo,GI,
Snooks,GF,BC,Romaine Br,Chan, *LH BC*

Gnaphalium uliginosum L.
  Weed n. to ~~NDB and~~ BofI. TC,GF,GeoP,
Tops,QVL,Trep, *Burg mig. BC*

G. supinum L.   (Gaspé)
  Englée (Waß).

G. sylvaticum L.
  BC,Pal. *BB*

G. norvegicum Gunn.   (Gaspé)
  Dock, On

Rudbeckia hirta L.
  StJ,Holyr,Tiddleton(Wag),GF,Meadows(Wag),
Steph,

Bidens frondosa L.
  GF,Meadows,BC,

Helianthus giganteus L.,var. subtuberosus (Bourgeau) Bn
  LH

Madia satva Broad Cove & Portugal Cove (Mus dye)

171

B. frond., var. stenodonta Fern.& StJ.
    Whit, NA,

Achillea Millefolium L. *Madman's Daisy*
    Weed, n. to NDB, and BofI. Tops, Torb, Gl,
    GF, GrandL, BC, WB, Trep, *Burg Mig*

A. Mil., var. nigrescens E. Meyer *Wild Tansy*
    Near coast s. to NDB and BStG. DR, Fogo,
    CH, BB, BC, FrenchmC, FC, *Burn Doct StP Pat*

A. Ptarmica L.
    StJ, HarbGrace (McGill), PS, BC, SpruceBr,
    Steph, *Burg*
    A. sibirica (Gaspé)
A. lanulosa Nutt.
    StJ,

Anthemis Cotula L.
    Clode's Sd, GF, FrenchmC (Wag),

A. arvensis L.   *A. tinctoria*
    Random (Wag),   *Gander etc. Murrays P (canal)*

Matricaria inodora L.
    Weed n. to NDB and BB. BB, GeoP, GF, StJ,
    StP,

M. grandiflora (Hook.) Britton   (Arct.)

M. Chamomilla L. *Bigin (canal)*
    Card, *Laakyard*

172.

M. suaveolens (Pursh) Buchenau
   StJ, Batw, BC. *Burg* BB

Chrysanthemum Leucanthemum L.
   EuP.

                                    Bachelor's Buttons
C. Leuc., var. pinnatifidum Lecoq&Lamotte
   Weed n. to NDB and BB. BB, Gl, MJ, GF, BC,
StG, *CB*

C. integrifolium Richardson   (Arct.)

Tanacetum vulgare L.
   Nat. n. to NDB and BofI. BC, FrenchmC,
Whit, *Lang* BB

T. huronense Nutt., var. terrae-novae Fern.
   PC, IB, GGard *MC. SC. Big. Cook, Norm. Boat, 4m.*
Sgg. Plc.

Artemisia canadensis Michx.
   BF, GF, MiddleArm, BlomBr, *Nv* BB

A. borealis Pall.
   Tab. GGard, *Sgg*
   a. bog var. latisecta Germ.
   Eg Nv BB
A. vulgaris L.
   StJ, BC.

A. Absinthium L.
   Tops, Kil, *EuP.*

[signature]
(German)

Tussilago Farfara L.
  StJ(Wag), BofI(Wag),
    (ama)

Petasites palmata (Ait.)Gray
  Freq. S. to NDB,Exploits and BStG. MaryAL,
TC, MJ, Fc, Big, Bum

P. vitifolia Greene   (Gaspé)

P. sagittata Pursh   (Lab.)

Arnica alpina (L.) Murr.   (Lab.)

A. plantaginea Pursh   (Lab.)

A. Sornbergeri Fern.   (Lab.)

A. terrae-novae Fern.
  GGard, Savc, Bum, Cook Pac,

A. pulchella Fern.
  Tab, Norm, Ha,

A. chionopappa Fern.
  Tab, GGard, Bum, SjS, Dvet, mid Pac Bb

A. gaspensis Fern.   (Gaspé)

A. Griscomi Fern.   (Gaspé)
    SjS,
  a. terrae a. Foerr
    Pac   Pac

A. mollis Hook.   (Gaspé)

Senecio vulgaris L.
   Com. weed.   StJ,FC,Chan,BarredI,Funk,.
BI,BC,CH,Trep, *North Burg Mis.*

S. sylvaticus L.
   Man,Kil,BJ,Whit,Cl,GF,Steph, *WS.*

S. Jacobaea L.
   StJ,BBulls,

S. palustris (D.) Hook.,v.congestus(RBr)
                        Hook.   (Lab.)

S. pauciflorus Pursh   (Gaspé)
     Nfd. acc. to B & B
   Ottis Mauve, Doct. On

S. indecorus Greene   (Gaspé)

S. resedifolius Less.
   Tab.

S. pauperculus Michx.
Com. on w. coast & e. to NDB. TC,FC,BC,BB
Fogo,Blom,CoalR,Holyr,MJ,Sum,MaryABr,JunctBr,
Chimn,GrandL,Blom,BP,Badg,BarredI,Fogo,GF,
RR,Gl,Marb,PaB,SC, *Burn, Quirp, On No. Han...
end.*

S. platensis Nutt (Gasp. Artic)

S. paup. var. firmifolius Greenm.
  IB, BB, SC. 599. On PL

S. aureus L.
  Gl, GF, PlacJ, BC, Tab, PaP, Salm, BenC. BB
S. pseudaureus Rydb.
  Tab BB
S. Pseudo-Arnica Less.
  Freq. on coast. FC, IB. On, Sit, Burg, Sit.

S. Rolandi Vict. (ming)

Arctium Lappa L.

A. minus (Hill) Bernh.
  ? StJ, Holyr. Gooseberry (aua)

a. nemorosum
  809 (aua)

Cirsium lanceolatum (L.) Hill
  Freq. n. to Trinity B and BofI. Man, BC, BB

C. muticum Michx. Horse Tops
  Com. PaP, StG, BB, GF, Blom, Brig, Doct, On ming.
     ♀ M
       BB

C. mut. var. monticola Fern.
  FC, IB, Blom s, Gov. No 68

C. arvense (L.) Scop.
  Com. n. to NDB and Bof I. GF, BC, BofI, StJ, BB
  mig.

C. palustre (L.) Scop.
  Humb, Howley

Carduus nutans L.
  Sit.

Centaurea nigra L. Broad Weed, French Clover
  Freq. weed n. to Exploits R and BofU. StJ, Tops,
  Humb, BC, FrenchmC, GeoP, BBulls, LH Sit. BB

  C. Cyanus, Exploits Stg
       (aua)

C. nigra, var. radiata DC.
   BC (E&G). *Sbo,*

Cichorium Intybus L.
   StJ, Botw, GF, *S&P,*

Hypochaeris radicata L.
   BC✓ *S&P, Murrays Pond (Acid)*

Leontodon autumnalis L. *Horse Dandelion, August Flower*
   Com. n. to NDB and BofI. StJ, GrandL,
Chan, FrenchmC, BC, *S&P, BB*

L. aut., var. pratensis (Link) Koch
   Com. n. to NDB and BB. Gl, BC, GrandL, BB, *Ha*
Quirp. Hán,

Taraxacum phymatocarpum Vahl   (Arct.)
   Burn

T. officinale Weber   *Dumbled'01* ✓
   Com. Balena, BC, GF, Torb, StB, StG, *Burg S&P,*
Ha'. Pac
T. palustre ✓
Pac vip.
T. latilobum DC.
   FC, StB, IB, GGard,

T. "arctolep"
Pac

T. lapponicum Kihlman ✓ ✓
   SB, SC, FC, GGard, *Doet, On Big Cook, Quirp, Burn,*
manure, 4-mi Pac

T. phymatocarpum Vahl.   *T. Heversm ✓*
                            Pac Doet
Burn.

T. graenlandicum *Dahlst.*
Cook, Burn, FC, SC, Bif, Ha, Pac pastures fields Ref (Aug) 177

T. ceratophorum (Ledeb.)DC.
SC,   Or Cook Ha, Big, Burn Pac

T. dymetorum *Greene*
Ha
T. lacerum Greene   T. foliosum
FC, Ha,   mid.

T.   T. nary g.   T. erythrospmum
Stg g.   mid, Pac.

Sonchus arvensis L.
Freq. n. to Exploits R. and BofI. Little
BellI, Salm, Tops, GF, BC, WC, Steph, StP, BB

S. oleraceus L.
Trep, StJ, Plac, Carb, Heart'sContent(Wag),
Random(Wag), StP, BB

S. asper (L.)Hill.
NHarb(W), Ol, GF, CoakR, Baie du Désespoir(Wag)
mig. BellI,(amd)

Lactuca spicata (Lam.)Hitchc.
Freq. n. to Expl.R and IB. GF, HawkeR,
WC, BC, HR, Doct, L Mt. BB

L. terrae-novae *Fern.*
Doct.
Agoseris gaspensis Fern.   (Gaspé)

Crepis biennis.
Kemp Bol (RBK)
Crepis nana Richardson (Lab.)
Burn.

Prenanthes racemosa Michx.   (Gaspé)

Pig Root (gives fine flav. to flesh)

P. trifoliolata (Cass.)Fern. *Mountain Turnip*
  Com. n. to NDB and BB. FrenchmC, BB, Carb,
Kil, StJ, GF, WB, Snooks, C, Torb, LarkH, BC, TC,
Tab *LH, Burg, Mig BB Hau*

P. trifol., var. nana (Bigel) Fern.
    Freq. on coast & mts. StJ, BarredI, CH, PR
FrenchmC, PS, Chan, FC, Brig, Trep, *On, Lmt, No*
H Florentinum
Hieb. aurantiacum
    Trinbrec (John McKeily)
Hieracium Pilosella L.
  StJ(wag), QV(Wil), BBulls, *Murrays Pond (Aua)*
    20-mile Pond (Aua)

H. vulgatum Fries.
  Whit, BC,

H. smolandicum Almg.
  Holyr, RockyR, StJ,

H. canadense Michx.
  Freq. Man, Tops GF, Torb, Snooks, HR, Steph,
SEA, MistC, StJ, Caultois, *On, Dvct, StB, Lu*
Mig, BW

H. can., var. hirtirameum Fern.
  WB, Cl, BF, GF, Tab. *Bear, Yank, Fd, BB Hau*

H. groenlandicum *Argeut - Tour.*
  Big, Burn, Quirp

Summary, June, 1925

	Indig.	Introd.		Indig.	Intr.

Polypodiac. 34 35   Salicac. 36 45 2
Schizeac. 1 1   Myricac. 3
Osmundac. 4 3   Betulac. 14
Ophioglossac. 5 1   Urticac. 1 3
Equisetac. 8 10   Santalac. 2
Lycopodiac. 21 23   Loranthac. 1
Selaginellac. 1 1   Polygonac. 14 11
Isoetaceae 4 4   Chenopod. 8 10 1
    Pteridophyta 78 84   Apeto. 79 14

Taxaceae 1 1   Caryoph. 29 38
Pinaceae 10 11   Portledac. 3
    Coniferales 11 12   Nymphaeac. 3
Typhaceae 1 1   Ranunc. 21 27
Sparganiac. 7 8   Berberid. 1
Najadac. 1 1   Fumariac 1 1
Potamogetonac 21 22   Cruciferae 26 23 06
Juncaginac. 3 3   Sarracen. 1
Alismac. 1 1   Droserac. 4
Gramineae 98 104 167   Crassul. 2 3
Cyperaceae 144 160   Saxifrag. 17 2
Araceae 1 1   Rosaceae 50 2
    1   Legum. 13 13
Eriocaulac. 1 1   Linac. 1 2
Xyridac. 1 1   Geran. 2 1
Juncac. 38 44   Euphorb. 1 3
Liliac. 10 11   Callitrich 4
Iridac. 6 6 11   Empetr. 3
Orchidac. 28 34   Aquifol. 23
    Monocots 361 17   Acerac. 3
    395   Balsamin 1

(over)

	Indig.	Introd.			Indig.	Introd.
Rhamn.	1		Plantagin.	5	3
Malvac.		2	Rubiac.	11 13	1
Hyperic.	4	1	Caprifol.	9	
Elatinac	1		Valerian.	1	
Cistaceae	2		Dipsacac		1
Violac.	10 12	2	Campanul	2	1
Elaeagn.	1		Lobeliac.	3	
Lythrac.		1	Composit.	86 10 35	
Onagrac	21 34		Gamopet.	237	70
Halorag.	5			265	80
Araliac.	2				
Umbellif	11 14	34	Pterid.	78	
Cornac	5		Conifer.	11	
Polypet	248	73	Monocots	450	17
Ericac.	43 46		Apets.	79	14
Diapens	1		Polypets	248	73
Plumbagin.	3		Gamopets	237	70
Primul.	9 10	34		1103	174
Oleac.	1			1014	
Gentianac.	8 9		Indig.	1103	
Apocynac.	3		Introd.	174	
Convolvul.	1 2			1277	
Boragin.	2	4 5		1188	
Labiatae	10	10 13			
Solanac.		1			
Scrophular	31 36	11			
Lentibul.	7				
Orobanch.	1				

Pterid 84
Conif 12
Monoc 395   22
apet,   99   14
Polyp. 293   82
Gamop 265   80
                ————————
Indiv   1148   198
Duplones 198
                ————————
        1346   (Sept, 1926)
Buenos Ayres   5   17
                ————————
        1153   215
        215
        ————
        1368

CPSIA information can be obtained
at www.ICGtesting.com
Printed in the USA
BVHW011918040522
636159BV00009B/180